OCR A LEVEL

MATHEMATICS A

EXAM PRACTICE

For Year 2

Heather Davis
Janet Dangerfield
Nick Geere
Rose Jewell
Sue Pope

HODDER
EDUCATION
AN HACHETTE UK COMPANY

Hachette UK's policy is to use papers that are natural, renewable and recyclable products and made from wood grown in well managed forests and other controlled sources. The logging and manufacturing processes are expected to conform to the environmental regulations of the country of origin.

Orders: please contact Hachette UK Distribution, Hely Hutchinson Centre, Milton Road, Didcot, Oxfordshire, OX11 7HH. Telephone: +44 (0)1235 827827. Email education@hachette.co.uk Lines are open from 9 a.m. to 5 p.m., Monday to Friday. You can also order through our website: www.hoddereducation.co.uk

ISBN: 978 1 5104 2368 8

© Heather Davis, Janet Dangerfield, Nick Geere, Rose Jewell, Sue Pope 2018

First published in 2018 by

Hodder Education,

An Hachette UK Company

Carmelite House

50 Victoria Embankment

London EC4Y 0DZ

www.hoddereducation.co.uk

Impression number 10 9 8 7 6 5 4 3 2 1

Year 2022 2021 2020 2019 2018

Cover photo © Denis Babenko/123RF.com

Typeset in Integra Software Services Pvt. Ltd., Pondicherry, India

Printed in the UK by Ashford Colour Press Ltd.

A catalogue record for this title is available from the British Library.

Contents

Full worked solutions and mark schemes are available online at www.hoddereducation.co.uk/OCRMathsExamPractice

Introduction

This book offers more than 300 questions to support successful preparation for the new A Levels in Mathematics. Grouped according to topic, the chapters follow the content of *OCR (A) A Level Mathematics Year 2*. Each chapter starts with short questions to support retrieval of content and straightforward application of skills learned during the course. The demand gradually builds through each chapter, with the later questions requiring significant mathematical thinking and including connections to other topics. This reflects the likely range of styles of question in the live papers. Answers are provided in this book, and worked solutions and mark schemes to all questions can be found online at **www.hoddereducation.co.uk/OCRMathsExamPractice.**

The data set referred to in some of the questions on statistics can be found at www.ocr.org.uk/qualifications/as-a-level-gce-mathematics-b-mei-h630-h640-from-2017/assessment/

1 Proof

1 Find counter-examples to disprove the following statements, where a and b are real numbers.

   (i) If $a < b$, then $a^2 < b^2$ [1 mark]

   (ii) If $a > b$, then $\dfrac{1}{a} < \dfrac{1}{b}$ [1 mark]

2 For the statements, **(i)-(iii)**, about A and B, choose the most appropriate option [a], [b] or [c].
[a] $A \Rightarrow B$
[b] $A \Leftrightarrow B$
[c] neither [a] nor [b]
In each case, justify your response.

   (i) A: $x \geqslant 0$; B: $|x| = x$ [2 marks]

   (ii) A: John is a pilot; B: John has good eyesight   [2 marks]

   (iii) A: a quadratic equation has two distinct roots; B: the discriminant of the quadratic equation is non-negative   [2 marks]

3 Comment on the correctness of the steps in the following argument.
$x - 2 = 0 \Rightarrow x(x - 2) = 0 \Rightarrow x = 0$ or $x = 2$;
but this is a contradiction of $x - 2 = 0$, as $0 - 2 \neq 0$.   [4 marks]

4 Prove, or disprove, the following statement.
'9 876 543 210 is a multiple of 6.'   [3 marks]

5 Prove that there is no positive integer, n, such that
$n^3 = n + 571\,423$.   [7 marks]

6 Derive the quadratic formula for the roots of the equation
$ax^2 + bx + c = 0$, by completing the square ($a \neq 0$).   [4 marks]

7 Prove that $\cos^2\theta + \sin^2\theta = 1$ when $0 < \theta < 90°$.   [4 marks]

8 Prove, or disprove, the following statement.
'If N is a positive integer, then $N(N + 2)$ is never a perfect square.' [3 marks]

9 Prove that there is just one pair of positive integers that
satisfy the equation $(x - 2)(y + 3) = 6$.   [5 marks]

10 **(i)** Show that, if $A \Rightarrow B$, then $B' \Rightarrow A'$ (where A' means
   'A is not true').   [3 marks]

   (ii) Show that, if $B' \Rightarrow A'$, then $A \Rightarrow B$.   [3 marks]

   (iii) Deduce that, to prove $A \Leftrightarrow B$, we can prove that $A \Rightarrow B$
   and $A' \Rightarrow B'$.   [4 marks]

2 Trigonometry

1 Solve the equation $\sin\theta = 0.25$ for $0° < \theta < 90°$. [1 mark]

2 Convert

   (i) 315° to radians [1 mark]

   (ii) $2\frac{4}{9}\pi$ radians to degrees. [1 mark]

3 A sector of a circle has arc length 20 cm and area 160 cm². Find the radius of the circle. [3 marks]

4 Write down the exact value of $\sqrt{8} \times \sin 315° \times \cos 30°$. [2 marks]

5 A sector of a circle of radius 8 cm has arc length 6 cm. Calculate the area of the sector. [3 marks]

6 A triangle has two sides of lengths 14 cm and 10 cm with an angle of 0.5 rad between them.
Calculate the area of the triangle. [2 marks]

7 A circle has radius 10 cm.
A segment of the circle is formed by a chord of length 12 cm.
Calculate the area of the segment. [4 marks]

8 Solve the equation $\cos\theta = 0.43$ for $-90° < \theta < 360°$. [3 marks]

9 **(i)** Show that $x = \frac{\pi}{4}$ is a solution of the equation

$$\sin x \times \cos x = \frac{1}{2}.$$ [2 marks]

   (ii) Write down another solution of this equation for which $-\pi < x \leqslant \pi$. [1 mark]

10 A sector of a circle of radius r cm has perimeter 12 cm.

   (i) Find an expression for the area of the sector in terms of r. [2 marks]

   (ii) Find the angle of the sector when the area is maximised. [3 marks]

11 Show that, for small positive angles, θ, in radians

   (i) $\dfrac{1 - \cos\theta}{\theta\sin 2\theta} \approx \dfrac{1}{4}$ [3 marks]

   (ii) $\lim\limits_{\theta \to 0}\left(\dfrac{1 - \cos 3\theta}{\theta\tan 3\theta}\right) \approx \dfrac{3}{2}$ [4 marks]

12 A series of photographs shows the Sun setting behind a pyramid. The bottom of each photograph is the horizon. Each picture is symmetrical about a vertical axis.

In each picture, the pyramid is shown as an isosceles triangle of base 6 cm and perpendicular height 4 cm.

The Sun is shown as the part of a circle of radius 5 cm that is above the horizon.

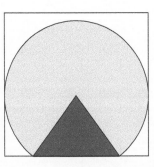 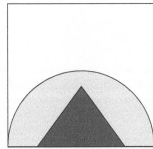 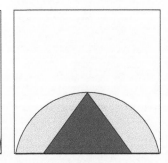

(i) In the first picture, the vertex of the triangle passes through the centre of the circle.
Calculate the area of the circle that is above the horizon and is not covered by the triangle. [2 marks]

(ii) In the second picture, exactly half the circle is above the horizon.
Calculate the area of the circle that is above the horizon and is not covered by the triangle. [2 marks]

(iii) In the third picture, the vertex of the triangle touches the circumference of the circle.
Calculate the area of the circle that is above the horizon and is not covered by the triangle. [3 marks]

In part (i) of this question you must show detailed reasoning.

13 (i) A circle of radius r cm is drawn with the largest possible regular hexagon that can be drawn *inside* the circle, and the smallest possible regular hexagon that can be drawn *outside* the circle, as shown.

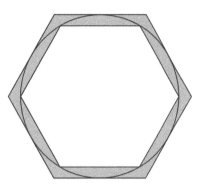

The region between the two hexagons is shaded.
Calculate the exact area of the shaded region. [6 marks]

(ii) Use the areas of the hexagons to deduce upper and lower bounds for π, giving your answers in surd form. [2 marks]

14 A crescent shape is made using two circles, as shown.

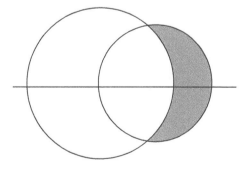

The smaller circle passes through the centre of the larger circle.

The larger circle has radius 10 cm.

The length of the shorter arc that forms the crescent is 15 cm.

The horizontal width of the crescent is 4 cm at its widest point.

Calculate, for the crescent shape,

(i) the perimeter [7 marks]

(ii) the area. [3 marks]

3 Sequences and series

1 Write down the first five terms of the sequence $u_n = 4 - 7n$. [1 mark]

2 Write down an expression for the nth term of the arithmetic sequence that starts with -2 and has a common difference of 1.5. [1 mark]

3 Write down the first five terms of the sequence $u_n = 2 \times 0.5^n$. [1 mark]

4 A geometric sequence has 2 and 4 as its third and fourth terms respectively. Find an expression for the nth term. [2 marks]

5 Show that the sum of the first one hundred counting numbers is 5050. [2 marks]

6 What is the condition for the geometric series $u_n = ar^n$ to converge? [1 mark]

7 Derive the formula for the infinite sum of a convergent geometric series. You may assume the formula for the sum of the first n terms of a geometric series. [3 marks]

8 Determine the minimum number of terms of the series $u_n = 3 \times 1.5^n$ for the sum to be greater than 100. [4 marks]

9 A geometric series with common ratio r, and an arithmetic series with common difference d, both start with 1.
The sums of their first ten terms are equal.

In this question you must show detailed reasoning.

(i) Show that $d = \dfrac{1}{9}\left(\dfrac{\left(r^{10} - 1\right)}{5(r - 1)} - 2\right)$. [4 marks]

(ii) Find d when $r = 2$. [2 marks]

10 Joan shares a cake between three people.
She cuts the cake into four equal pieces and gives each person a piece.

She divides the remaining piece into four equal pieces and gives each person a piece.

She continues this process indefinitely.

(i) Show that, assuming there are no crumbs, this process would give each person a third of the cake. [4 marks]

(ii) Joan wants to share a cake between n people.
Using the same process as before, how many pieces should Joan cut the cake into? Justify your answer. [3 marks]

11 The sum of an arithmetic series whose terms are all integers, is zero.

(i) Express the first term a, in terms of the common difference d and the number of terms n. [3 marks]

(ii) Which of the following statements could be true in this context? Justify your answers.

(a) a and d are both positive. [1 mark]

(b) a is negative and d is a positive multiple of 2. [1 mark]

(c) a is positive, d is negative and n is odd. [1 mark]

(d) a is negative, d is positive and n is even. [1 mark]

12 The picture shows a decreasing sequence of squares within squares that continues indefinitely.
The side length of each unshaded square is half the side length of the previous unshaded square.

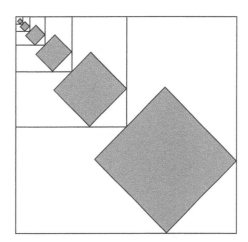

(i) Show that the area of the largest shaded square is $\frac{2}{9}$ of the area of the largest square. [3 marks]

(ii) What fraction of the largest square is shaded altogether? [3 marks]

13 The first, sixth and twenty-sixth terms of an arithmetic sequence $(d \neq 0)$ are three consecutive terms of a geometric sequence.
What is the common ratio of the geometric sequence? [7 marks]

14 The sum of the first and third terms of a geometric sequence is six times the second term of the geometric sequence.

Find, in exact form, the possible values of the common ratio r. [5 marks]

4 Functions

1 Describe each of the following mappings, A, B and C, as either one-to-one, one-to-many, many-to-one, or many-to-many, and state whether it represents a function. [3 marks]

A   B   C

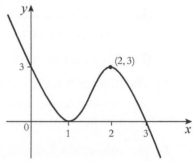 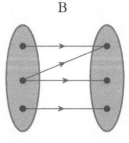 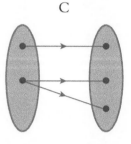

2 Find the range of the function $f(x) = x^2 - 5$. [1 mark]

3 This graph shows a function $y = f(x)$.

Sketch the function $y = f(x + 1)$, indicating the coordinates of any points that you know. [3 marks]

4 The functions f and g are defined by $f(x) = 3x - 7$ and $g(x) = x^2 + 1$. Find $gf(x)$ as a quadratic function of x. [2 marks]

5 The functions f and g are defined by $f(x) = 3x - 7$ and $g(x) = x^2 + 1$. Solve the equation $fg(x) = 14$. [3 marks]

6 Write the inequality, $2 < x < 5$, in the form $|x - a| < b$, where a and b are numbers to be found. [2 marks]

7 Sketch the graph of $y = |2x - 1|$. [2 marks]

8 (i) Find the range of the function $f(x) = 2(x + 2)^2 - 5$ defined on the domain $x \geqslant 0$. [2 marks]

   (ii) Find the range of the function $g(x) = 2(x - 2)^2 + 5$ defined on the domain $x \geqslant 0$. [2 marks]

   (iii) For what values of x is $g(x) > f(x)$? [3 marks]

9 The curve $y = x^2(2x - 3)$ has a minimum turning point at $(1, -1)$.

   (i) (a) The curve is translated so that the minimum turning point is at the origin.

       Find the equation of the translated curve. [2 marks]

       (b) The translated curve is then reflected in the x-axis so that the minimum turning point becomes a maximum turning point.

       Find the equation of this reflected curve. [2 marks]

(ii) (a) Explain why the function $y = x^2(2x - 3)$ does not have an inverse function. [1 mark]

  (b) Give a suitable domain on which $y = x^2(2x - 3)$ does have an inverse function. [1 mark]

10 The graph of a continuous function $y = f(x)$ passes through the points $(0,3)$, $(2,1)$ and $(5,-2)$.

  (i) What coordinates are known on each of the graphs of each of the following functions?

  (a) $y = f(x - 2)$ [1 mark]

  (b) $y = 3f(x)$ [1 mark]

  (c) $y = f(x) - 1$ [1 mark]

  (d) $y = f(2x)$ [1 mark]

  (ii) What can be deduced about $y = f^{-1}(x)$ if the graph of $y = f(x)$ also passes through each of the following points?

  (a) $(3,3)$ [1 mark]

  (b) $(3,-1)$ [2 marks]

11 It is given that $f(x) = 5x + 2$, $g(x) = 3x^2 + 1$ and $h(x) = 2^x$ for $x > 0$.

  (i) (a) Calculate the value of $fgh(1)$. [1 mark]

  (b) Find the range of the function $fgh(x)$. [2 marks]

  (ii) (a) Find and simplify algebraic expressions for $fg(x)$ and $gf(x)$. [3 marks]

  (b) Hence solve the equation $fg(x) = gf(x)$. [2 marks]

12 The graph shows $y = f(x)$ where $f(x) = \dfrac{x - a}{x - ka}$ for some positive constants a and k, where $k > 1$.

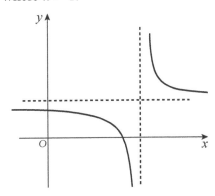

  (i) Where does the graph cross the axes? [2 marks]

  (ii) Find an expression for $f^{-1}(x)$. [2 marks]

  (iii) In the case when $a = 2$ and $k = 2$, solve the equation
  $$f^{-1}(x) = \frac{1}{f(x)}.$$ [3 marks]

13 (i) Sketch the graphs of $y = |x + 2|$ and $y = |2x - 4|$ on the same axes. [3 marks]

  (ii) Solve the inequality $|x + 2| > |2x - 4|$. [4 marks]

5 Differentiation

1 What is the condition for a point of inflection on the function
   $y = f(x)$? [1 mark]

2 It is given that $\dfrac{d^2y}{dx^2} < 0$ at a stationary point.

   Describe the nature of the stationary point. [1 mark]

3 Under what circumstances is a point of inflection a stationary
   point? [1 mark]

4 For what values of x is the function $f(x) = x^3 - 12x - 4$ **not**
   increasing? [3 marks]

5 Use the chain rule to differentiate $y = \sqrt{3x + \dfrac{1}{x}}$. [3 marks]

6 Use the product rule to differentiate $y = x\sqrt{x - 5}$. [4 marks]

7 Use the quotient rule to differentiate $y = \dfrac{x}{x^2 + 3x - 2}$. [3 marks]

8 Find the points of inflection of the curve
   $y = x^4 - 2x^3 - 12x^2 + 2x - 1$. [5 marks]

9 A graph is drawn of the function $f(x) = x^3 - x + 10$.

   (i) For what values of x will the curve be convex? [3 marks]

   (ii) Find the coordinates of the point of inflection. [2 marks]

10 A pond is modelled as a cone with its radius and height in the ratio 2:3.
   Snow melt is filling the pond at a rate of $7.5\,\text{cm}^3$ per second.

   (i) Determine the rate at which the depth of water is increasing at
   the point when the radius of the pond is 50 cm. [5 marks]

   (ii) Suggest a way in which the model may have to change as the pond
   gets deeper. [1 mark]

11 The curve $x = y + \dfrac{3}{2y^2 + 1}$ is shown here.

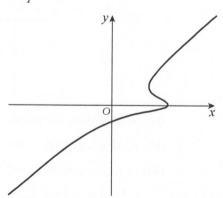

   Find the gradient of the curve where it crosses the y-axis. [6 marks]

12 Water is being poured into a cylindrical tank of radius 2 m at
   the rate of $6\,\text{m}^3\,\text{min}^{-1}$.
   Find the rate at which the depth of the water, h, in the tank
   is rising. [5 marks]

13 The function $y = 2x^4 + ax^2 + 2x + b$ has a point of inflection at $\left(\dfrac{1}{2}, 1\right)$.
   Find the coordinates of its other point of inflection. [7 marks]

14 Air is escaping from a spherical balloon at a rate of $2\,\text{cm}^3\text{s}^{-1}$.
When the radius of the balloon is $10\,\text{cm}$, find the rate of decrease of

    (i) the radius   [3 marks]

    (ii) the surface area.   [4 marks]

In this question you must show detailed reasoning.

15 Prove that, on any cubic curve, the x coordinate of the point of inflection is midway between the x coordinates of the two stationary points.   [7 marks]

16 A function is given by $y = x + 2 - \dfrac{4}{x + 1}$.

    (i) Prove that y increases as x increases.   [3 marks]

    (ii) Find the coordinates of the two points on the graph of the function where the gradient is 2.   [4 marks]

    (iii) The tangents at these points cut the x-axis at P and Q.

         Find the length PQ.   [3 marks]

6 Trigonometric functions

1 A triangle has sides of lengths $3\,\text{cm}$, $5\,\text{cm}$ and $7\,\text{cm}$.
Calculate, to the nearest 0.1 rad, the smallest angle in the triangle. [2 marks]

2 Solve the equation $\tan 2x = 1$ for $90° \leqslant x < 180°$. [2 marks]

3 By considering appropriate graphs, find the number of solutions of the equation
$2x = \tan x$ in the interval $-2\pi < x < 2\pi$. [3 marks]

4 Write down the exact value of $\operatorname{cosec} \dfrac{\pi}{6}$. [2 marks]

5 Find the roots of the quadratic equation
$2\cos^2 x - 13\cos x + 6 = 0$
for which $-\pi \leqslant x \leqslant \pi$. [3 marks]

6 If $\cos x = \dfrac{1}{4}$, where x is an acute angle, find the exact value of $\tan x$. [2 marks]

7 Solve the equation $\cot x = 2\cos x$ in the range $-\pi < x < \pi$. [4 marks]

8 Solve the equation $2\cos^2 x - 3\cos x + 1 = 0$ for $0 \leqslant x < \dfrac{\pi}{2}$. [3 marks]

9 Solve the equation $\sin(3x - \dfrac{\pi}{6}) = 1$, in the range $-\pi < x \leqslant \pi$. [4 marks]

10 (i) Sketch the graphs $y = \arcsin x$ and $y = \arctan x$ on the same axes, where the angles are measured in radians. [2 marks]

   (ii) (a) Write down the exact values, in radians, of $\arcsin(1)$ and $\arctan(1)$. [2 marks]

      (b) Show that $\dfrac{\arcsin(1)}{\arctan(1)}$ is not the same as $\arccos(1)$. [2 marks]

11 Find the possible value, or values, of $\operatorname{cosec} x$ if $\cot x = \sec x$, giving your answers in the form $\dfrac{a + \sqrt{b}}{c}$ where a, b and c are integers. [6 marks]

12 A pyramid has perpendicular height $h\,\text{m}$ and is built on a square base of edge length $2x\,\text{m}$.

A person whose eye level is at a height of $1.60\,\text{m}$ stands at a point C at the midpoint of an edge of the base. They walk away from the pyramid in a direction that is perpendicular to the edge of the base.

When the person has walked $50\,\text{m}$, they reach a point B.

The person then turns and looks up at E, the vertex of the pyramid. The angle of elevation of E above the person's eye level is $40°$.

The person walks a further $50\,\text{m}$ away from the pyramid, along the same line as before, to reach a point A that is $100\,\text{m}$ from C.

The person again turns and looks up at E. The angle of elevation of E above the person's eye level is now $30°$.

   (i) By using the sine rule on the triangle ABE, calculate the length BE. [3 marks]

   (ii) Calculate

      (a) the height of the pyramid [2 marks]

      (b) the edge length of the square base. [2 marks]

13 (i) Show that $\dfrac{\sec^2\theta - 2}{\cos\theta + \sin\theta} \equiv \sec\theta(\tan\theta - 1)$, provided
$\cos\theta + \sin\theta \neq 0$. [3 marks]

(ii) Hence solve the equation
$\dfrac{\sec^2\theta - 2}{\cos\theta + \sin\theta} = \tan\theta - \sec^2\theta$ for $0 \leqslant \theta \leqslant 2\pi$. [5 marks]

14 (i) Show, algebraically, that if $\csc^2 x = 2\cot x$, then $\cot x = 1$. [2 marks]

(ii) Hence solve the equation $\csc^2(4\theta + \pi) = 2\cot(4\theta + \pi)$ for
$-\dfrac{\pi}{2} \leqslant \theta \leqslant \dfrac{\pi}{2}$. [4 marks]

(iii) Deduce the number of solutions of the equation
$\csc^2(5\theta)\tan(5\theta) = 2$ in the interval $0 \leqslant \theta \leqslant \dfrac{\pi}{2}$. [2 marks]

7 Further algebra

1 For what values of x is the binomial expansion of $(1 + 3x)^{-1}(1 - 5x)^{-2}$ valid? [2 marks]

2 Show that $\dfrac{1}{\sqrt{9 + 6x}} = \dfrac{1}{3}\left(1 + \dfrac{2x}{3}\right)^{-\frac{1}{2}}$. [3 marks]

3 Write the fraction $\dfrac{x^3 + 3x^2 + x}{x - 1}$ in the form

$Ax^2 + Bx + C + \dfrac{D}{x - 1}$. [2 marks]

4 Write the first four terms of the binomial expansion of $(1 + x)^{-2}$ in ascending powers of x. [3 marks]

5 Simplify the algebraic fraction $\dfrac{x^3 + 2x^2 + x}{2x^4 + 2x^3 - x - 1}$. [3 marks]

6 By substituting $x = \dfrac{1}{50}$ in the expansion $(1 - x)^{\frac{1}{2}} \approx 1 - \dfrac{1}{2}x - \dfrac{1}{8}x^2$, find a rational approximation to $\sqrt{2}$. [3 marks]

7 Find the expansion of $(1 + 3x)(1 + x)^{-2}$ as far as the term in x^3. For what values of x is this expansion valid? [4 marks]

8 Express $\dfrac{2x + 5}{(1 - 3x)(3 + x)^2}$ in the form $\dfrac{A}{1 - 3x} + \dfrac{B}{3 + x} + \dfrac{C}{(3 + x)^2}$ where A, B and C are constants to be found. [5 marks]

9 (i) Write the first three terms of the binomial expansion of $(1 + 2x)^{\frac{1}{3}}$ in ascending powers of x. [4 marks]

(ii) For what values of x is this expansion valid? [1 mark]

In this question you must show detailed reasoning.

10 Let $f(x) = \dfrac{4x^3 - x + 6}{2x^2 + x - 1}$.

Show that $f(x)$ can be written in the form $Ax + B + \dfrac{C}{mx + n} + \dfrac{D}{px + q}$ where A, B, C, D and m, n, p, q are constants to be found. [5 marks]

11 Find the binomial expansion of $(8 - x)^{-\frac{1}{3}}$ as far as the term in x^2. [5 marks]

12 (i) Write the expression $\dfrac{3x + 1}{x^2 - 1} + \dfrac{4}{x - 1}$ as a single fraction in its simplest form. [2 marks]

(ii) Given that $(1 - x^2)^{-1} = 1 + x^2 + x^4 + \dots$, find a cubic approximation to $\dfrac{3x + 1}{x^2 - 1} + \dfrac{4}{x - 1}$ for small values of $|x|$. [2 marks]

(iii) Calculate the absolute percentage error in your approximation from part (ii) when $x = 0.1$. [2 marks]

13 (i) Find a quadratic approximation for $\dfrac{(6 - x)(1 + 2x)}{1 + x}$. [2 marks]

(ii) State the values of x for which the approximation is valid. [1 mark]

(iii) For what positive values of x is the absolute error in the approximation smaller than $\dfrac{x}{1 + x}$? [4 marks]

14 (i) Write $\dfrac{4x + 5}{(1 + 5x)(1 - x)}$ as a sum of partial fractions with constant numerators. [3 marks]

(ii) Find the binomial expansion of $\dfrac{4x + 5}{(1 + 5x)(1 - x)}$ up to and including the term in x^3. [5 marks]

(iii) For what range of values is the expansion in part (ii) valid? [1 mark]

8 Trigonometric identities

1 Use the result for $\sin(A + B)$ to prove that $\sin 2\theta = 2\sin\theta\cos\theta$. [2 marks]

2 Use the result for $\cos(A + B)$ to prove that $\cos 2\theta = 2\cos^2\theta - 1$. [3 marks]

3 Find the coordinates of the maximum value of $y = R\cos(x - \alpha)$, where $0 \leqslant \alpha \leqslant 180°$. [2 marks]

4 Describe the transformations that map the graph of $y = \cos x$ onto the graph of $y = R\cos(x - \alpha)$. [3 marks]

5 Solve $\tan 3\theta = \tan\theta$ for $0 \leqslant \theta \leqslant \pi$. [4 marks]

6 Determine the exact value of $\sin 75° - \sin 15°$. [4 marks]

7 Show that $\cos 4\theta = 8\cos^4\theta - 8\cos^2\theta + 1$. [3 marks]

8 (i) Express $\cos x - \sin x$ in the form $R\cos(x + \alpha)$ where $0 \leqslant \alpha \leqslant 180°$. [4 marks]

(ii) Determine the minimum value of $\cos x - \sin x$. [1 mark]

(iii) Solve $\cos x - \sin x = 0.1$ for $0 \leqslant x \leqslant 360°$. [4 marks]

9 (i) Express $2\sin x - 3\cos x$ in the form $R\sin(x - \alpha)$, where $0 < x < \dfrac{\pi}{2}$ and R is given in exact form. [4 marks]

(ii) A function, $f(x) = \dfrac{1}{k + 2\sin x - 3\cos x}$, is defined on the interval $0 < x < 2\pi$.
Find the coordinates of the minimum value in $0 < x < 2\pi$, in terms of k. [3 marks]

(iii) The minimum value is $\dfrac{1}{2}$.
Find k. [2 marks]

10 The value of $\tan\theta = \dfrac{a}{b}$.
Find the value of $a\cos 2\theta - b\sin 2\theta$. [6 marks]

11 Derive an exact value for $\tan 15°$ from the exact value of $\tan 30°$. [6 marks]

12 Solve $\cos 2\theta - \sin\theta - 1 = 0$, for $0 \leqslant \theta \leqslant 2\pi$. [6 marks]

13 (i) Express $\sqrt{3}\cos x + \sin x$ in the form $R\cos(x - \alpha)$. [4 marks]

(ii) Hence, or otherwise, solve
$\sqrt{3}\cos x + \sin x = \sqrt{2}, 0° \leqslant x \leqslant 360°$. [3 marks]

14 Solve $\cos\left(\theta + \dfrac{\pi}{6}\right) = 2\sin\left(\theta + \dfrac{\pi}{3}\right)$, $-\pi \leqslant \theta \leqslant \pi$. [9 marks]

9 Further differentiation

1 Prove that the derivative of the function $y = \ln x$ is $\dfrac{1}{x}$. [3 marks]

2 The function $y = \sin kx$ has a local maximum when $x = \dfrac{\pi}{6}$.
Determine the value of k. [3 marks]

3 Find the gradient of the curve $y = \ln 2x$ at the point it cuts the x-axis. [3 marks]

4 Show that the x coordinates of the stationary points of the function $y = x \cos x$ must satisfy the equation $x = \cot x$. [4 marks]

5 Find an expression for the gradient function of the hyperbola $ax^2 - by^2 = c$. [3 marks]

6 The function $f(x) = ke^{1-x^2}$ has a maximum value of 1. Determine the value of k. [3 marks]

7 Find the derivative of $y = \tan^{-1} 3x$. [4 marks]

8 Find the derivative of $y = \sin^{-1} x$. [4 marks]

9 (i) Show that the function $f(x) = \sin x - k \tan x$ only has stationary values when $|k| \leq 1$. [3 marks]

(ii) When $k = \dfrac{1}{8}$, find all the stationary values for $0 \leq x \leq 2\pi$. [2 marks]

10 This graph shows the function $y = x \ln x$.

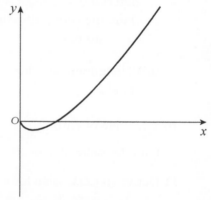

Find the minimum value of the function. [6 marks]

11 This graph shows the ellipse $3x^2 + 4y^2 = 12$.

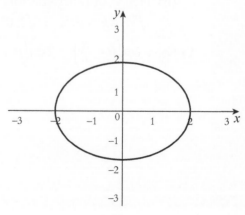

(i) Find the equations of the tangents to the ellipse when $x = 1$. [6 marks]
(ii) Show that these tangents meet on the x-axis. [2 marks]

12 A circle has equation $x^2 + y^2 = r^2$.

   (i) Find the equation of the normal to the circle at the point on the circle (a,b).   [6 marks]

   (ii) Show that the normal passes through the centre of the circle.   [1 mark]

13 The function $y = x^2 \ln(x + 1)$ describes the path of a particle.
At time t, the x coordinate of the position of the particle is $x = t^2 - 2t$.
Find the rate of change of the y coordinate of the position of the particle, with respect to time, when $t = 3$.   [6 marks]

In this question you must show detailed reasoning.

14 The path of a point is described by the function $y = e^x \sin 3x$.
At time t, the y coordinate is given by $y = \cos 2t$.
Show that x is stationary with respect to time when $t = 2\pi$.   [6 marks]

10 Integration

For all the definite integrals, use your calculator to check your answers. You must show all your working to gain full credit.

1 Find an exact value for $\int_0^1 5e^{3x}\,dx$. [3 marks]

2 Find $\int (\cos 3x + 2\sin 3x)\,dx$. [3 marks]

3 Evaluate $\int_0^{\frac{\pi}{6}} \sec^2 2t\,dt$, giving your answer in terms of $\sqrt{3}$. [3 marks]

4 Find $\int \left(2t - 3t^{\frac{2}{3}}\right) dt$. [3 marks]

5 Find $\int \dfrac{1}{5 - 2t}\,dt$. [3 marks]

6 Use integration by parts to show that $\int \ln x\,dx = x\ln x - x + c$. [3 marks]

7 It is given that $f'(x) = x(3 + x^2)^4$.
Given that $f(0) = 24$, find an expression for $f(x)$. [5 marks]

8 Find $\int \dfrac{3t + 6}{t(t + 3)}\,dt$, giving your answer in the form $\ln(f(t))$. [6 marks]

In this question you must show detailed reasoning.

9 Using a suitable substitution, or otherwise, show that
$$\int_0^1 (5x + 3)^5\,dx = \frac{52\,283}{6}.$$ [5 marks]

10 Find the area of the shaded region in the graph below, which is bounded by the curves $y = x^3 + 2x^2 - 3x + 1$ and $y = 2x^2 + x + 1$. [6 marks]

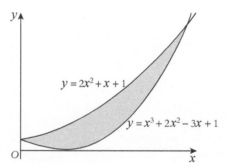

11 Find $\int \dfrac{1}{x} \ln x\,dx$. [3 marks]

12 Using a suitable substitution, or otherwise, show that
$$\int_0^{\frac{\pi}{6}} \frac{\cos 3x}{4 + 2\sin 3x}\,dx = \frac{1}{6}\ln\left(\frac{3}{2}\right).$$ [6 marks]

13 Find the total area of the shaded regions in the graph below, between the curve $y = 4\sqrt{x} - x^2\sqrt{x}$, the x-axis and the line $x = 4$.

Give your answer in the form $p + q\sqrt{2}$, where p and q are rational.

[7 marks]

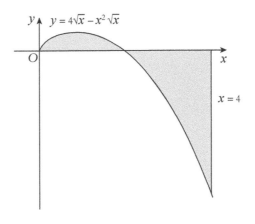

14 Find $\int \left(e^x + 2\right)^2 \, \mathrm{d}x$. [4 marks]

15 Use a trigonometrical identity to find $\int 6\sin^2 2x \, \mathrm{d}x$. [4 marks]

16 (i) Find the exact value of $\int_0^2 5xe^{x^2} \, \mathrm{d}x$. [5 marks]

(ii) Find the exact value of $\int_0^2 5xe^{2x} \, \mathrm{d}x$. [5 marks]

17 Evaluate $\int_0^{\frac{\pi}{2}} 12\sin x \cos x \, \mathrm{d}x$. [5 marks]

18 Show that $\int_{\frac{\pi}{4}}^{\frac{\pi}{2}} \cot x \, \mathrm{d}x = \int_{\frac{1}{\sqrt{2}}}^1 \left(\frac{1}{u}\right) \, \mathrm{d}u$, where $u = \sin x$.

Evaluate the integral, giving your answer in the form $k\ln 2$. [6 marks]

19 Below is the graph of $y = 6x\sqrt{1 - 2x}$.

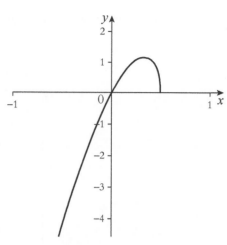

(i) Show that the substitution $u = 1 - 2x$ transforms the integral
$$\int 6x\sqrt{1 - 2x}\,dx \text{ into } \frac{3}{2}\int\left(u^{\frac{3}{2}} - u^{\frac{1}{2}}\right)du$$
[3 marks]

(ii) Hence find the exact area of the region bounded by the curve and the x-axis. [5 marks]

20 Find $\displaystyle\int \frac{3x^2 - x}{(3x + 1)^2\,(x + 1)}\,dx$. [8 marks]

21 (i) Use the substitution $u = \ln x$ to find $\displaystyle\int \frac{1}{x\ln x}\,dx$. [4 marks]

(ii) Hence find an exact value in its simplest form for
$$\int_e^{e^2} \frac{1}{x\ln x}\,dx.$$
[3 marks]

22 Below is part of the graph of $y = x^2 \sin x$.

Find the exact area of the shaded region. [8 marks]

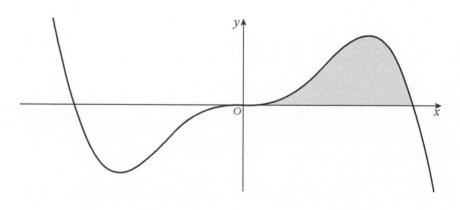

11 Parametric equations

1 A curve has the parametric equations $x = 3t^2$, $y = 6t$.
   Find the Cartesian equation of the curve. [2 marks]

2 A curve has parametric equations $x = 3t^2$, $y = 6t$.
   Find $\dfrac{\mathrm{d}y}{\mathrm{d}x}$. [3 marks]

3 Convert the parametric equations $x = 2t$, $y = \dfrac{2}{t}$ into Cartesian form. [2 marks]

4 Write down the parametric equations for a circle, with its centre at the origin and a radius of 2. [2 marks]

5 State parametric equations for a circle with its centre at (1,2) and a radius of 1. [2 marks]

6 A circle has parametric equations $x = \cos\theta$, $y = \sin\theta$.
   Write down the values of θ that correspond to the points on the y-axis. [2 marks]

7 A circle has parametric equations $x = \cos\theta$, $y = \sin\theta$.
   Find the value(s) of θ for which $\dfrac{\mathrm{d}y}{\mathrm{d}x} = 1$. [4 marks]

8 A curve has parametric equations $x = t^2$, $y = t^2 + 1$.
   Why is it **not** correct to say that the curve is the straight line $y = x + 1$? [1 mark]

9 The curve below has parametric equations $x = t^3 + \dfrac{1}{2}t^2$, $y = 2t$.

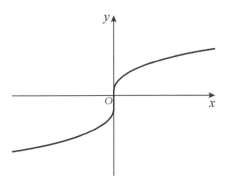

   (i) Determine the Cartesian equation of the curve, such that all its coefficients are the smallest possible integers. [2 marks]

   (ii) Use the Cartesian form to find the points at which $\dfrac{\mathrm{d}y}{\mathrm{d}x} = 1$. [5 marks]

   (iii) Determine $\dfrac{\mathrm{d}y}{\mathrm{d}x}$ in terms of t. [2 marks]

   (iv) Hence find the values of t for which $\dfrac{\mathrm{d}y}{\mathrm{d}x} = 1$. [3 marks]

10 A spot moves over the screen of an oscilloscope such that, at time t, its position is given by $x = \cos 3t$ and $y = \sin 2t$.

   (i) Find an expression for the square of its speed. [4 marks]

   (ii) Show that this is a minimum when $t = 2\pi$. [6 marks]

11 A curve is defined by the parametric equations $x = \tan\theta$, $y = \cos\theta$.

(i) Find the Cartesian equation of the curve. [2 marks]

In this question you must show detailed reasoning.

(ii) Find the coordinates of the stationary points of the curve, for $0 \leqslant \theta \leqslant \pi$. [10 marks]

12 A particle moves such that, at time t, $x = \cos 2t + \sin t$ and $y = e^{-t}$.
Find the rate of change of y, with respect to x, when $t = \pi$. [6 marks]

13 A curve is defined by the parametric equations
$x = 3\cos\theta$, $y = 2\sin\theta$ $(0 \leqslant \theta < 2\pi)$

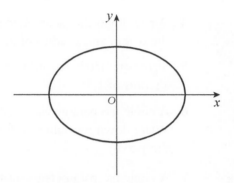

(i) Find the Cartesian equation of the curve. [2 marks]

(ii) The normals to the curve, at certain points, pass through the point (1,0).
Find the coordinates of those points. [12 marks]

14 A curve is defined by the parametric equations $x = t + \dfrac{1}{t}$, $y = t - \dfrac{1}{t}$.

Find the Cartesian equation of the curve. [3 marks]

15 A curve is defined by the parametric equations $x = \dfrac{t}{3-t}$, $y = \dfrac{t^2}{3-t}$.

Find the Cartesian equation of the curve, in the form $y = \mathrm{f}(x)$. [5 marks]

12 Vectors

1 Find the magnitude and direction of the vector $\mathbf{a} = 10\mathbf{i} + 24\mathbf{j}$. [4 marks]

2 A force, F, acts at an angle of $30°$ above the negative x-axis, and has magnitude 10 N.
Express F in component form. [3 marks]

3 Determine the magnitude of the vector $\mathbf{b} = \begin{pmatrix} -1 \\ 2 \\ -2 \end{pmatrix}$. [2 marks]

4 Find the distance between the points with position vectors
$\begin{pmatrix} 3 \\ -1 \\ -4 \end{pmatrix}$ and $\begin{pmatrix} -5 \\ 6 \\ 3 \end{pmatrix}$. [2 marks]

5 Find a unit vector parallel to the line joining $(1,-2,3)$ and $(-3,1,-2)$. [3 marks]

In this question you must show detailed reasoning.

6 Use vectors to show that the midpoints of the sides of the quadrilateral ABCD form the vertices of the parallelogram PQRS. [8 marks]

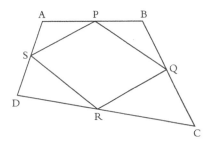

7 Two forces act on an object, as shown in the diagram.

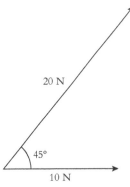

Find the magnitude and direction of the resultant force. [8 marks]

8 It is given that the resultant of the three vectors, $\begin{pmatrix} a \\ 3 \end{pmatrix}$, $\begin{pmatrix} 2 \\ b \end{pmatrix}$ and $\begin{pmatrix} 1 \\ 4 \end{pmatrix}$,
has magnitude $5\sqrt{2}$, and is in a direction of $45°$ to the positive x-axis.
Find a and b. [8 marks]

9 This diagram shows a regular hexagon.

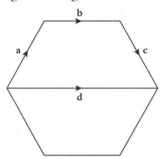

Express **c** and **d** in terms of **a** and **b**. [5 marks]

10 The points A and B have coordinates (−2,4) and (1,2) respectively.
The point C divides AB in the ratio 1:4.

Use a vector method to determine the coordinates of C. [4 marks]

13 Differential equations

1 Give brief definitions of the following terms.

    (i) First order differential equation. [1 mark]

    (ii) General solution of a differential equation. [1 mark]

    (iii) Particular solution of a differential equation. [1 mark]

2 Show that $y = \dfrac{1}{4 - x}$ is a solution of the differential equation
$\dfrac{dy}{dx} = y^2$, where $y = 1$ when $x = 3$. [3 marks]

3 Find the general solution of the differential equation
$\dfrac{dy}{dx} = 2x - 1$. [2 marks]

4 Find the general solution of the differential equation $\dfrac{dy}{dx} = xy$. [4 marks]

5 Find the particular solution of the differential equation
$\dfrac{dy}{dx} = 2x - 1$ that satisfies $x = 1$, $y = 1$. [2 marks]

6 Find the equation of the curve that satisfies the differential equation
$\dfrac{dy}{dx} = xy$ and passes through the point $(2,1)$. [3 marks]

7 Solve the differential equation $\dfrac{dy}{dx} = 2y - 1$, given that $y = 3$
when $x = 0$. [6 marks]

8 The acceleration of a car is known to be inversely proportional to
its velocity.

    (i) Write down a differential equation to model this situation. [2 marks]

    (ii) Refine this model, given that it is known that the
acceleration is $2\,\mathrm{m\,s^{-2}}$ when the velocity is $10\,\mathrm{m\,s^{-1}}$. [2 marks]

    (iii) Find the general solution of the differential equation. [5 marks]

    (iv) Find the particular solution of the differential equation, given
that the car is initially stationary. [1 mark]

    (v) At what time will the car be travelling at $20\,\mathrm{m\,s^{-1}}$? [2 marks]

9 Food in an oven heats up at a rate that is proportional to the difference
between the temperature of the oven and the temperature of the food.
The oven is at a constant temperature of $175\,^{\circ}\mathrm{C}$.

  Food at $15\,^{\circ}\mathrm{C}$ is placed in the oven.

  The food initially heats up at a rate of $0.5\,^{\circ}\mathrm{C\,s^{-1}}$.

  Find a differential equation to model this situation. [4 marks]

10 The world's population is modelled by the differential equation $\dfrac{dP}{dt} = kP$,
where P is the number of people in billions and t is the time in years.
Given that P was estimated to be 7.5 at the end of 2017, and that it is
estimated to be 12 at the end of 2100, use the model to predict the world's
population at the end of 2200 (to the nearest billion). [9 marks]

11 Solve the differential equation $x\dfrac{dy}{dx} = y(2 - y)$, for $x > 0$ and $0 < y < 2$,
given that $x = 1$ when $y = 1$. [10 marks]

14 Numerical methods

1 Show that the equation $x^5 - x^2 - 3 = 0$ has a root between $x = 1$ and $x = 2$. [2 marks]

2 The curves $y = e^{x^2}$ and $y = 4 - x^3$ intersect at the points where $x = a$ and $x = b$.

Show that the x coordinates of one of these points lies between 0.9 and 1.1.

State the accuracy of the answer estimated at this stage. [3 marks]

3 The function $f(x) = e^x - \dfrac{1}{x - \sqrt{2}}$ crosses the x-axis between $x = 1.6$ and $x = 1.7$.

Explain why a decimal search method might fail to find it. [1 mark]

4 The trapezium rule is used to estimate the shaded area in this graph.

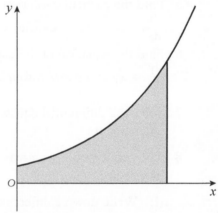

Is the estimate an underestimate or overestimate?
Justify your answer. [1 mark]

5 The value of $\displaystyle\int_0^4 \sqrt{x^2 + 5x + 1}$ is to be estimated.

Calculate an upper and lower bound for the integral, using four rectangular strips. [3 marks]

6 The graph shows the function $f(x)$.

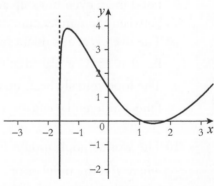

Explain why the Newton-Raphson method may fail to find all of the roots of $f(x) = 0$. [1 mark]

7 The equation $x^5 + x^2 - 4 = 0$ has a solution in the interval [1,2].

(i) Show that $x = \sqrt[5]{4 - x^2}$ is a rearrangement of the equation.   [1 mark]

(ii) This spreadsheet shows the first steps in solving the equation using the rearrangement in part (i) as an iterative formula.

	A	B
1	x	g(x)
2	1	1.245731
3	1.245731	1.196101
4	1.196101	1.207715
5	1.207715	1.205079
6	1.205079	1.205681
7	1.205681	1.205544
8	1.205544	1.205575
9	1.205575	1.205568

(a) What starting value is used?   [1 mark]

(b) What formula is used in cell A3?   [1 mark]

(c) From the information on the spreadsheet, give the root to an accuracy of 4 decimal places and check that this is valid.   [2 marks]

8 This spreadsheet is used to calculate the value of an integral using the trapezium rule with six strips.

	A	B	C	D
1		x	y	
2	0.5	0	0	
3		0.5	2.125	0.53125
4		1	3.4	1.38125
5		1.5	3.825	1.80625
6		2	3.4	1.80625
7		2.5	2.125	1.38125
8		3	0	0.53125
9				7.4375

(i) What is the width of each strip?   [1 mark]

(ii) State the formula in cell D5.   [1 mark]

(iii) Write down the estimated value of the integral.   [1 mark]

9 Sam is using the iterative formula

$$x_{n+1} = \frac{2x_n^3 - 4x_n^2 - 2}{3x_n^2 - 8x_n + 2}$$

to find a root of the equation $x^3 - 4x^2 + 2x + 2 = 0$.

Sam uses a spreadsheet and declares, after the iterations shown, that the root must be 1.3111 to 4 decimal places.

	A	B
1	x	
2	1	1.333333
3	1.333333	1.311111
4	1.311111	1.311108

Is Sam correct?

Is Sam justified in making that assertion?

Justify your answers.   [3 marks]

10 The equation $\ln(x^2 + 1) - x^3 - 0.01 = 0$ has three distinct roots.

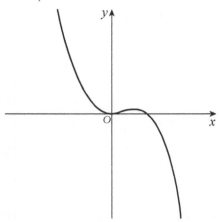

(i) Show that one of the roots lies between $x = 0.1$ and $x = 0.2$. [2 marks]

(ii) Using a change of sign method, determine this root to an accuracy of 3 decimal places, verifying that it is accurate to 3 decimal places. [3 marks]

(iii) For each of the other roots, determine intervals of length 0.1, in which they lie. [2 marks]

(iv) Explain clearly why a decimal search method may not succeed in finding all of the roots of this equation. [2 marks]

11 The function f(x) is defined by f(x) = $xe^x - 1$.

(i) Show that there is a root in the interval [0,1]. [2 marks]

(ii) Show that the equation $x = e^{-x}$ has the same solution as f(x) = 0. [2 marks]

(iii) Using the iterative formulae $x_{n+1} = e^{-x_n}$ and $x_0 = 0$, determine the root to an accuracy of 2 decimal places. [4 marks]

(iv) Sketch the functions f(x) = x and g(x) = e^{-x} and show the first four iterations on your graph. [3 marks]

(v) Explain why the formulae in part (iii) took many iterations to converge sufficiently to write down the root to 2 decimal places. [1 mark]

(vi) If you started with $x_0 = 1$, would you have got to the answer in part (iii) more quickly?
Justify your answer. [1 mark]

12 (i) Use the trapezium rule with three strips to estimate the value of $\int_2^5 \frac{1}{x-1}\,dx$.

You should show details of your method. [3 marks]

(ii) Establish, using a graph, whether your estimate is an overestimate or an underestimate. [2 marks]

13 The function $f(x)$ is defined by $f(x) = x^5 - 3x + 1$.

    (i) Show that the equation $f(x) = 0$ has a root in the interval $[-1, -2]$.   [2 marks]

    (ii) Use the Newton-Raphson formula to show that an iteration formula to find the roots of $f(x) = 0$ may be written as

$$x_{n+1} = \frac{4x_n^{\,5} - 1}{5x_n^{\,4} - 3}.$$ [3 marks]

    (iii) Find all of the roots to the equation $f(x) = 0$ accurate to 2 decimal places.   [6 marks]

14 The height of the cross section of a tunnel at intervals of $2\,m$ are shown in the table.

You may assume that the walls of the tunnel are vertical for $3\,m$ at either side.

Distance (m)	0	2	4	6	8	10	12	14	16
Height (m)	3	7.8	9.6	10.4	10.7	10.4	9.6	7.8	3

    (i) Use the trapezium rule with three ordinates to estimate the area of the cross-section of the tunnel.   [2 marks]

    (ii) Use the trapezium rule with nine ordinates to estimate the area of the cross-section of the tunnel.   [2 marks]

    (iii) Which is the better estimate? Justify your answer.   [1 mark]

15 The curve $y = e^{\frac{1}{2}x} - 2x$ intersects the x-axis at two points.

    (i) Show that one of these points is in the interval $[0, 1]$ and find the interval with consecutive integer endpoints that contains the other point of intersection.   [3 marks]

    (ii) Show that the equation $e^{\frac{1}{2}x} - 2x = 0$ can be rearranged into the following forms.

      ① $x = \frac{1}{2} e^{\frac{1}{2}x}$

      ② $x = 2\ln 2x$ [2 marks]

    (iii) Use these rearrangements to perform iterations to find the two roots to an accuracy of 1 decimal place.   [4 marks]

    (iv) Show, on a diagram, the successive iterations that find the root in the interval $[0, 1]$.   [2 marks]

    (v) Explain, in detail, why both iterative formulae are needed.   [4 marks]

16 Use the Newton-Raphson formula to find the solution to the equation $x^2 e^{-x} - 0.5 = 0$ to 2 decimal places.   [7 marks]

17 The velocity of a particle, at time t, is given by $v = 3t\cos^2 0.5t$.

    (i) Using four rectangles in each case, find a lower bound and an upper bound for the distance travelled, in metres, by the particle in the first π seconds.   [3 marks]

    (ii) Use the trapezium rule with five ordinates to estimate the distance travelled, in metres, in the first π seconds.   [2 marks]

    (iii) How could the estimate in part **(ii)** be improved?   [1 mark]

18 The equation $x^4 - 3x^2 - x + 2 = 0$ has two roots.

    (i) Show that the equation can be rearranged to give the iterative formula
$$x_{n+1} = \sqrt{\frac{1}{3}\left(x_n^4 - x_n + 2\right)}.$$
  [2 marks]

    (ii) Find the root between 0 and 1, to an accuracy of 2 decimal places, using this formula.   [2 marks]

    (iii) Show, using a diagram, how this formula fails to find the other root that is in $[1,2]$.   [2 marks]

    (iv) Use the iterative formula $x_{n+1} = \sqrt{\dfrac{3x_n^2 + x_n - 2}{x_n^2}}$ to determine the other root to an accuracy of 2 decimal places.   [2 marks]

15 Probability

1 Two fair, six-sided dice are thrown together.
  D is the event that both dice show the same number.
  S is the event that the total of the two scores is seven.
  Show that S and D are mutually exclusive. [2 marks]

2 The events A and B are independent.
  $P(A) = 0.6$ and $P(B) = 0.7$.
  Find $P(A \cup B)$. [3 marks]

3 Three children play a game in which they take it in turns to throw
  a ring over a peg.
  Xena goes first and has a probability of success of 0.5, Yolanda goes second
  with a probability of success of 0.4, and Zac goes last with a probability of
  success of p.
  Calculate the value of p, given that the probability Zac is the first
  to win, on his first attempt, is 0.195. [3 marks]

4 Two events, A and B, have probabilities p and $2p$ respectively.
  The conditional probability $P(A|B) = \frac{p}{3}$.
  Find an expression in terms of p for the probabilities
  (i) $P(A \cap B)$ [2 marks]
  (ii) $P(A \cup B)$. [2 marks]

5 This Venn diagram shows the probabilities for two events, A and B,
  and their intersections.

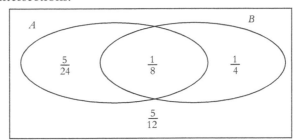

Determine whether the events are independent, mutually exclusive
or neither. [4 marks]

6 The tree diagram shows the probability of meeting a red light at two sets of
  traffic lights on the same road.

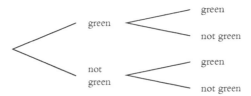

The probability of the first set being green is 0.4.

When the first set is green, the probability of the second set not being
green is 0.3.

When the first set is not green, the second set is equally likely to be
green or not.

Complete the tree diagram and calculate the probability of meeting a
green light at the second set. [4 marks]

7 The two-way table shows the number of Year 11 students that study French and German at a school.

Which is more likely, that a French student is male, or that a male student studies French? [5 marks]

	French	German	Total
Male			170
Female		83	
Total	165		320

8 Events A and B are mutually exclusive.
Events A and C are also mutually exclusive.
Events B and C are independent.
You are given that $P(B) = \frac{1}{2}$, $P(C) = \frac{1}{3}$ and $P(A \cup B \cup C)' = \frac{1}{12}$.
Draw a Venn diagram and calculate $P(A)$. [6 marks]

9 A college has 1090 students.
635 of the students are female.
455 of the female students have long hair.
53 of the male students have long hair.

(i) Draw a Venn diagram showing the two sets
F = {female students} and L = {students having long hair}. [3 marks]

(ii) A student is chosen at random.
Calculate the probabilities of choosing

(a) a male student [2 marks]

(b) a female student with long hair [1 mark]

(c) a female student, given that the student has long hair [2 marks]

(d) a student with long hair, given that the student is female. [2 marks]

In this question you must show detailed reasoning.

(iii) Show that the events 'being female' and 'having long hair'
are not independent. [3 marks]

10 A football coach selects players for her squad by setting three tests which, experience shows, are independent.
A player is selected if they pass at least two of the three tests.

Test A has a 90% pass rate, test B has an 80% pass rate, and test C has a 75% pass rate.

(i) Calculate the probability that a randomly chosen player passes two out of the three tests. [3 marks]

(ii) Calculate the expected number of players to be selected from 50 players who have a trial. [4 marks]

(iii) Calculate the probability that a player selected for the squad has passed all three tests. [2 marks]

11 An art gallery trials a new piece of equipment that claims to be able to recognise a fake picture with 95% accuracy.
The gallery owners believe that only one in 200 of their paintings are fake.

   (i) Complete the tree diagram.   [2 marks]

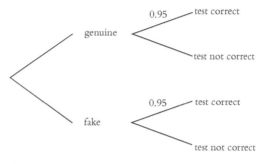

   (ii) Calculate the probability that a picture will be recognised as fake.   [3 marks]

   (iii) Calculate the probability that a picture *recognised* as fake is *actually* fake.   [3 marks]

16 Statistical distributions

1 The random variable X is normally distributed with mean 75 and variance 10.

   (i) Write down the standard deviation of X.   [1 mark]

   (ii) Calculate the probability that

      (a) $X > 85$ [2 marks]

      (b) $X < 80$ [2 marks]

      (c) $70 < X < 78$.   [2 marks]

2 The weight of soap powder packed into boxes is normally distributed with mean $1002\,g$ and standard deviation $1.2\,g$.

   (i) Calculate the probability that the weight of powder exceeds $1005\,g$.   [2 marks]

   (ii) The weight printed on the box is $1\,kg$.
What proportion of boxes are underweight?   [2 marks]

3 $X \sim N\left(45,\ 10^2\right)$
Given that $P(X > k) = 0.7$, find the value of k, giving your answer to 3 significant figures.   [4 marks]

4 The speed of cars on a stretch of motorway has a mean of $68\,mph$ with a standard deviation of $5\,mph$.
One day, the fastest 8% of drivers were fined by the police.

At what speed did the police choose to fine drivers? You can assume that the speeds of cars are normally distributed.   [4 marks]

5 $X \sim N(\mu,\ 17^2)$
Given that $P(X < 100) = 0.25$, find the value of μ, giving your answer to 4 significant figures.   [4 marks]

6 $X \sim N(35,\ \sigma^2)$
Given that $P(X > 36) = 0.4$, find the value of σ, giving your answer to 4 significant figures.   [4 marks]

7 A random variable X has a normal distribution $X \sim N(20, 3^2)$.
Sketch a graph showing this distribution.

Your sketch should show the position of the points of inflection.

Shade the regions of the graph for which the values of X would be outliers.   [4 marks]

8 A random variable X has a normal distribution with mean 5 and standard deviation 1.2.
Another random variable Y is defined by $Y = 3X - 1$.

   (i) Write down the distribution of Y.   [3 marks]

   (ii) Calculate the probability that Y exceeds 10.   [2 marks]

9 The distance that children in a PE class throw a beanbag is normally distributed with mean $4.8\,m$ and standard deviation $0.45\,m$.
The teacher gives gold stars to children who throw more than $5.4\,m$, silver stars to children who throw more than $4\,m$ but less than $5.4\,m$, and 'I tried' stickers to the other children.

For a class of 30 children, estimate how many of each colour star and stickers she awards. [6 marks]

10 The mass of lettuces sold by a supermarket is normally distributed with mean 560 g and standard deviation 20 g.

(i) Find the probability that a randomly chosen lettuce weighs more than 595 g. [2 marks]

(ii) A customer buys three lettuces.
Calculate the probability that each of them weighs more than 595 g. [2 marks]

(iii) How many lettuces in a box of 24 would be expected to weigh less than 530 g? [3 marks]

11 A doctor uses the formula $w = 1.1x - 100$ to estimate the weight of a child, w, in kg from their height, x, in cm.
He assumes the heights of five year-olds are normally distributed with a mean of 108 cm and a standard deviation of 2.5 cm.

(i) Explain why the doctor believes the mean weight of five year-olds is 18.8 kg and find the standard deviation. [3 marks]

(ii) Use these values to estimate the 90th percentile for weight (the height below which 90% of the population lie).
Give your answer in kilograms to 1 decimal place. [3 marks]

12 Bashira has researched information about the size of leaves of the chestnut oak tree and finds that the leaves are between 10 cm and 22 cm long.
She assumes the leaf size fits a normal distribution.

(i) Explain why Bahira estimates the standard deviation to be approximately 2 cm. [2 marks]

(ii) Write down an estimate for the mean. [1 mark]

(iii) Sketch a graph showing the distribution of lengths, and shade the area of the graph that represents the probability that a randomly chosen leaf is less than 15 cm long. [3 marks]

(iv) Using these values and a normal distribution, calculate the probability that a randomly chosen leaf is less than 15 cm long. [2 marks]

13 A machine cuts lengths of cable from a roll.
It is set so that the lengths of the pieces are normally distributed with mean μ and standard deviation 0.5 cm.

The machine produces cables that are supposed to be 2 m long.

(i) What mean must be set so that 95% of pieces are at least 2 m long?
Give your answer to 3 decimal places. [4 marks]

The cut lengths are packed in boxes of 100.

(ii) How many of the pieces, in each box, would be expected to be more than 202 cm long? [3 marks]

14 A company producing car tyres advertises with the information that 90% of its tyres will last at least 23 000 km.
Elsewhere in the information, the company states that the average distance before its tyres will need replacing is 28 500 km.

(i)   Calculate the standard deviation for the life of a tyre.   [4 marks]

(ii) Calculate the probability that a tyre will last 35 000 km.   [2 marks]

15 Eggs are sold in four sizes.
The table shows the weight range for each size.

Each size band starts at the minimum weight and includes eggs up to, but not including, the maximum weight.

Assume 25% of eggs in the UK fall into each category.

Size	Weight (g)
very large	73 and over
large	63–73
medium	53–63
small	53 and under

(i)   Find the mean and standard deviation of eggs in the UK.   [4 marks]

(ii) Determine the ranges of values of weight for an egg to be considered an outlier.   [3 marks]

16 A teacher believes about 3% of students achieved a Grade 9 in their GCSE, with the lowest Grade 9 achieved with 248 marks out of 300. She also believes that 20% of students achieved a Grade 7 or above, with the lowest Grade 7 achieved with 159 marks.

She assumes that the marks fit a normal distribution with mean μ and standard deviation σ.

She uses the information about Grade 9 to obtain the equation
$\mu + 1.881\sigma = 248.5$.

(i)   Show how the teacher obtained this equation.   [3 marks]

(ii) Find a second, similar, equation connecting μ and σ.   [2 marks]

(iii) By solving the equations in parts (i) and (ii) simultaneously, find the values of μ and σ.
Give your answers to 4 significant figures.   [2 marks]

(iv) Show that these values indicate that about 15% of students would obtain a mark less than zero.   [2 marks]

(v) What does this figure indicate about the assumptions that the teacher has made?   [2 marks]

Hypothesis testing

1 Explain the key feature of a scatter diagram that indicates that rank correlation, and not the correlation coefficient, should be used.   [1 mark]

2 A hypothesis test is carried out using the value of rank correlation. Which word should appear in the conclusion of the test: *correlation* or *association*?   [1 mark]

3 This scatter diagram shows the body surface area (BSA) of a sample of 25 patients plotted against their mass:

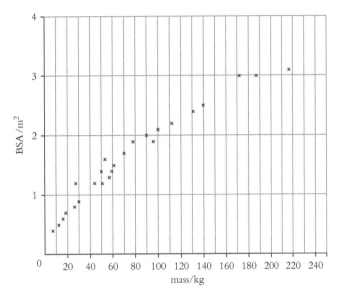

(i) State the feature of this scatter diagram that indicates that rank correlation should be used.   [1 mark]

(ii) The graph-drawing software gives the value of 0.9863 for the rank correlation coefficient.
The critical value for this sample size for a 2-tail hypothesis test at the 5% level is 0.3977.
Perform the hypothesis test, stating your hypotheses and conclusions clearly.   [5 marks]

4 This scatter diagram shows the ages of 11 people and their time spent watching television in the last week:

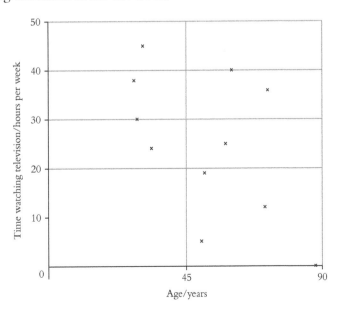

The graph-drawing software states that the rank correlation coefficient is −0.2727.

A hypothesis test is to be carried out to see if there is any association between age and the time spent watching television.

(i) Write down the null and alternative hypotheses for this test. [2 marks]

(ii) The critical value for a 2-tail test at the 5% level for the rank correlation coefficient is 0.5874.
Complete the test, stating your conclusions carefully. [3 marks]

5 This scatter diagram shows the height, in cm, and age, in years, of a sample of 55 pine trees:

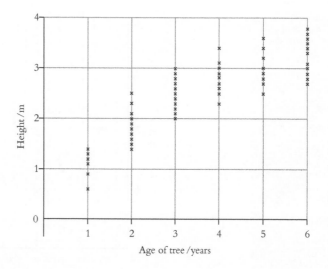

(i) Explain why the scatter diagram appears to be in vertical stripes. [1 mark]

(ii) What feature of the scatter diagram indicates that rank correlation should be used as the basis of a hypothesis test? [1 mark]

(iii) The graph-drawing software gives the value 0.8603 for the rank correlation coefficient.
The critical value for a 1-tail test at the 1% level is 0.3139.

Perform a hypothesis test to see if there is a positive association between the age and height of the trees. [5 marks]

6 Maria sells ice cream from a van in the park every day in September. This scatter diagram shows her sales and the maximum air temperature on each day:

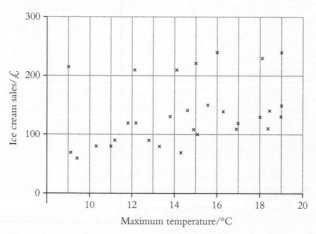

(i) Suggest a reason why the scatter diagram seems to have two distinct groups of points. [1 mark]

(ii) The correlation coefficient for this set of data is 0.3814.
The critical value for a 2-tail test at the 5% level is 0.3610.

Carry out a hypothesis test to see if there is evidence of any correlation between ice cream sales and maximum air temperature. [5 marks]

(iii) Maria claims there is proof of association between ice cream sales and maximum air temperature.
Make two comments about her claim. [2 marks]

7 A hypothesis test is to be carried out to test whether the mean of a normal distribution is 14.
The null hypothesis can be written as $H_0 : \mu = 14$.
Write down the three possible forms for the alternative hypothesis.
[3 marks]

8 A random variable has a normal distribution $N(\mu, \sigma^2)$.
A sample of size n is taken from the population.
Write down the distribution of the sample mean $\overline{X}$. [3 marks]

9 A sample of size 10 is taken from a population that is normally distributed with standard deviation 2.
The null hypothesis is $H_0 : \mu = 50$ and the test is carried out at the 10% significance level.

(i) Write down the distribution of the sample mean $\overline{X}$ under the null hypothesis. [2 marks]

(ii) Find the critical region for each of these possible alternative hypotheses.

    (a) $H_1 : \mu > 50$ [2 marks]

    (b) $H_1 : \mu < 50$ [2 marks]

    (c) $H_1 : \mu \neq 50$ [2 marks]

10 A population is normally distributed with mean 25 and standard deviation 8.
A sample of size 100 is taken and the sample mean $\overline{X}$ found.
Calculate the probabilities that

(i) $\overline{X} < 24$ [2 marks]

(ii) $\overline{X} < 23$ [2 marks]

(iii) $\overline{X} > 26.5$. [2 marks]

11 An athletics club introduces a new training regime to reduce 100 m sprint times for its athletes.
The sprint times were normally distributed with mean 13.5 s and standard deviation 1.5 s.

After training, a sample of five athletes have a mean time of 12.5 s.

Assume that the standard deviation has not changed.

The coach performs a hypothesis test at the 5% significance level to see if the sprint times have reduced.

(i) Write down the distribution of $\overline{X}$, the sample mean. [1 mark]

(ii) Write down the null and alternative hypotheses. [2 marks]

(iii) Calculate the probability that the sample mean is less than 12.5 s. [2 marks]

(iv) Complete the hypothesis test, stating your conclusions clearly. [3 marks]

12 A farmer believes the heights of wheat plants on his farm are normally distributed with mean 130 cm and standard deviation 10 cm. He wants to test, at the 1% significance level, if growing wheat near to the road reduces the height of the plants.

A sample of 20 plants grown near the road have a mean of 125.5 cm.

(i) Write down the null and alternative hypotheses that the farmer should use. [2 marks]

(ii) Find the critical region for this test. [3 marks]

(iii) Complete the test, stating your conclusions clearly. [3 marks]

13 Arina weighs a sample of eight goldfinches on a remote island to see if they are significantly different from European goldfinches, whose weights are normally distributed with mean 17 g and standard deviation 1 g. She uses $H_0 : \mu = 17$ as her null hypothesis.

(i) Write down the alternative hypothesis Arina should use. [1 mark]

(ii) Arina conducts her test at the 5% significance level. Find the critical region for her test. [3 marks]

Arina's results, in grams, are 17, 18, 19.5, 15.5, 16, 16, 19, 18.

(iii) Complete the hypothesis test, stating the conclusion clearly. [4 marks]

14 A teacher thinks that the attention span of the children in her class has a mean of 21 minutes with a standard deviation of 3 minutes. She decides to test whether playing classical music in class makes any difference to the children's attention span.

She measures the attention span of a sample of 10 children, while playing classical music, and finds the mean time is 23 minutes.

(i) Write down the null and alternative hypotheses for her test. [2 marks]

(ii) Write down the distribution of $\overline{X}$, the sample mean. [2 marks]

(iii) Complete the hypothesis test, stating your conclusions clearly. [4 marks]

15 A golf club keeps records of the scores of its members and finds they are normally distributed with mean 87.3 and standard deviation 9. A sample of 20 club members using a new type of putter have a mean score of 85.4.

Test, at the 5% significance level, whether the new type of putter has reduced their scores significantly. Assume that the standard deviation has not changed. [8 marks]

18 Kinematics

Use $g = 9.8\,\text{m}\,\text{s}^{-2}$ unless indicated otherwise.

1 A ball is shot vertically upwards.
   It rises through a vertical height of 2.5 metres before coming to rest.
   What was the initial upward speed of the ball?　　　　　　　[2 marks]

2 A train consists of an engine and eight carriages, each of length 21 m.
   The train accelerates from rest.

   Rob is sitting on a seat on the platform. When the train starts to move,
   Rob is alongside the front of the train. It takes 5 seconds for the engine
   to pass Rob.

   The train stops accelerating when it reaches a speed of $28\,\text{m}\,\text{s}^{-1}$.

   How long after starting does the train finish passing Rob?　　[6 marks]

3 A particle starts at the origin and moves with velocity $v = 4 - t^2$, where t
   is the time in seconds from when the particle starts to move.
   The particle stops when it returns to the origin.

   How long does it take for the particle to return to the origin?　[4 marks]

4 A stone is thrown from the top of a cliff with initial velocity $2\mathbf{i} + 0.3\mathbf{j}$.
   It moves under gravity and lands in the sea after 1.635 seconds in flight.

   Let the initial position of the stone be $0\mathbf{i} + Y\mathbf{j}$ and the final position
   be $X\mathbf{i} + 0\mathbf{j}$.

   (i) Find the values of X and Y.　　　　　　　　　　[3 marks]

   (ii) After t seconds of flight ($0 < t < 1.635$), the stone is at position
   $x\mathbf{i} + y\mathbf{j}$.
   By eliminating t, find a quadratic function that describes the
   trajectory of the stone.　　　　　　　　　　　　　　[3 marks]

   (iii) How high is the stone above the sea when it is travelling at 45°
   to the horizontal?　　　　　　　　　　　　　　　　[3 marks]

5 At midday, a ship has position $3\mathbf{i} + 11\mathbf{j}$, measured from a coastguard
   station, where $\mathbf{i}$ is a unit vector due east and $\mathbf{j}$ is a unit vector due north.
   The ship travels with constant velocity $4\mathbf{i} - 3\mathbf{j}$ for 2 hours.

   (i) Calculate the position of the ship after 2 hours.　　　[2 marks]

   (ii) When is the ship closest to the coastguard station?　　[3 marks]

6 A particle starts at rest at the point $(2,3,5)$.
   The acceleration at time t seconds from the start is given by

   $$\mathbf{a} = 6(t^2 - 1)\mathbf{i} + \frac{1}{25}t^4\mathbf{j} - 4t\mathbf{k}.$$

   Find the values of x and y when the position of the particle
   is $x\mathbf{i} + y\mathbf{j} - 13\mathbf{k}$.　　　　　　　　　　　　　[8 marks]

Forces and motion

1   A box of weight 20 N sits on a slope inclined at 10° above the horizontal.
    Find the component of the weight down the slope.   [2 marks]

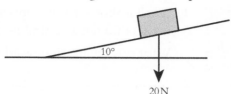

2   A boy uses a light, inextensible rope to pull a crate of weight 250 N across a
    rough horizontal floor.
    The rope makes an angle of 30° with the horizontal.

    Draw and label a force diagram to show the four forces acting on
    the crate.   [3 marks]

3   A particle is in equilibrium under the action of forces
    $\mathbf{F}_1 = 9\mathbf{i} + 3\mathbf{j}$, $\mathbf{F}_2 = -2\mathbf{i}$, $\mathbf{F}_3 = \mathbf{i} - 7\mathbf{j}$, and $\mathbf{F}_4$

    where $\mathbf{i}$ and $\mathbf{j}$ are unit vectors acting horizontally and vertically.
    Find the force $\mathbf{F}_4$.   [2 marks]

4   A force is $-4\mathbf{i} + 7\mathbf{j}$ N in component form.
    Find the magnitude of the force and the angle between the force
    and the direction of the vector $\mathbf{i}$.   [2 marks]

5   A resultant force of $\begin{pmatrix} 13 \\ 4 \end{pmatrix}$ N acts on a particle of mass 0.5 kg.

    Find, in vector form, the acceleration of the particle.   [2marks]

6   Force $\mathbf{P}$ has magnitude 6 N and acts along a bearing of 080°.
    Force $\mathbf{Q}$ has magnitude 4 N and acts along a bearing of 180°.
    Find the magnitude of the resultant force $\mathbf{P} + \mathbf{Q}$. [4 marks]

7   Box A, of mass 2 kg, sits on a smooth horizontal table.
    A light, inextensible string connects box A to box B. Box B has a mass
    of 1 kg.

    The string passes over a small, smooth pulley at the edge of the table.

    The string between box A and the pulley is horizontal, and the string
    between the pulley and box B is vertical.

    The system is released from rest.

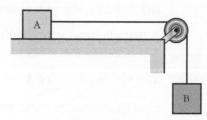

    Write down the equation of motion for box A and the equation
    of motion for box B.   [2 marks]

8 A block of weight 40 N sits at rest in equilibrium on a rough slope that is inclined at 10° to the horizontal.

   (i) Draw a force diagram to show the forces acting on the block. [1 mark]

   (ii) State the direction in which the frictional force acts, and calculate its magnitude. [2 marks]

   The block is then pulled up the slope using a string that is parallel to the slope.
   The frictional force is now 9 N and the acceleration of the block is $1\,\text{m}\,\text{s}^{-2}$.

   (iii) Calculate the tension in the string. [3 marks]

9 A sack of weight 240 N is suspended from a small, smooth pulley at B, through which a light, inextensible rope ABC passes.
   The rope is attached to a horizontal beam at A and C.

   The horizontal distance between A and C is 3.6 m and the length of the rope ABC is 4.5 m.

   (i) Calculate $\cos\theta$, where θ is the angle BAC. [2 marks]

   (ii) Calculate the tension in the rope. [2 marks]

10 A sack of weight 240 N is tied to two light, inextensible ropes: AB, of length 2.16 m, and BC, of length 2.88 m. The distance AC is 3.6 m.

   (i) Show that the ropes are perpendicular. [1 mark]

   (ii) Find the value of

       (a) cos BAC [1 mark]

       (b) cos BCA. [1 mark]

   (iii) Find the tension in

       (a) rope AB [3 marks]

       (b) rope BC. [1 mark]

11 A tractor of mass 3000 kg pulls a loaded trailer of mass 1500 kg.
   The total resistance on the tractor is 800 N and the total resistance on the trailer is 2000 N.

   (i) Write down the equation of motion for the trailer and hence calculate the force in the tow-bar when the tractor is moving horizontally at a steady speed. [2 marks]

   (ii) Find the drive force from the tractor's engine when the tractor is moving horizontally at a steady speed. [2 marks]

   The tractor starts to climb a hill that is inclined at θ to the horizontal, where $\sin\theta = 0.02$.
   The acceleration of the tractor is $0.01\,\text{m}\,\text{s}^{-2}$.

   (iii) Calculate the drive force. [3 marks]

12 Forces $\mathbf{F}_1 = 6\mathbf{i} + 7\mathbf{j}$ N, $\mathbf{F}_2 = x\mathbf{i} - 5\mathbf{j}$ N and $\mathbf{F}_3 = 4\mathbf{j}$ N act on a particle of mass m kg to give it an acceleration of $2\mathbf{j}\,\text{m}\,\text{s}^{-2}$.

   (i) Find the value of x. [3 marks]

   (ii) Find the value of m. [1 mark]

13 A goods train, consisting of an engine of mass 70 tonnes and five trucks each of mass 20 tonnes, is travelling at a constant speed of $12\,\mathrm{m\,s^{-1}}$ on level horizontal ground.

The resistance on each truck is $500\,\mathrm{N}$.

The last truck becomes uncoupled (disconnected).

The driving force and resistances on the engine and remaining trucks are unchanged.

(i)   What is the new acceleration of the train?   [2 marks]

When the train has been accelerating for 3 seconds, the other trucks become uncoupled from the engine, although they are still coupled to each other.

(ii) How far do these trucks travel from when they become uncoupled until when they come to rest?   [4 marks]

14 The resultant of two forces, P and Q, acting on a particle, is R.
Force Q has magnitude 2P and force R has magnitude 1.4P.

(i)   Calculate the angle between the forces P and Q.   [3 marks]

(ii) Find the magnitude of the resultant of P and −Q.   [3 marks]

(iii) Calculate the angle between P + Q and P − Q.   [3 marks]

20 Moments

1 The (positive) forces X, Y and Z are applied to the light rod AC, as shown in this diagram.

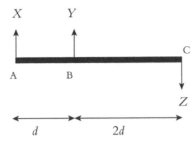

Explain why the rod cannot be in equilibrium. [3 marks]

2 In general terms, how can three equations be created if a solid object is at rest? [3 marks]

3 If the forces on an object are balanced, how would you decide about which point to take moments? [3 marks]

4 Three vertical forces, of X N, 30 N and 10 N, are applied to a light rod of length 1 m, as shown in the diagram below.
The force of X N is applied at a distance of d m from the left-hand end of the rod, and the force of 30 N is applied at the mid-point of the rod.

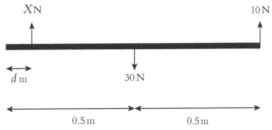

What values must X and d have in order for the rod to be in equilibrium? [5 marks]

5 The (positive) forces X, Y and Z are applied to the light rod AC, as shown in the diagram below.

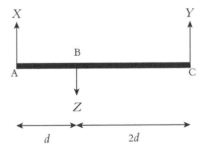

Assuming that the rod is in vertical equilibrium, show that the total moments about A, B and C are equal. [8 marks]

6 Forces are applied to the light rod AC, as shown in the diagram below.

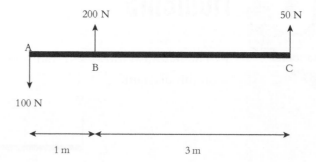

Find the magnitude and line of action of the additional force that would be needed in order for the rod to be in equilibrium. [7 marks]

21 Projectiles

Use $g = 9.8\,\mathrm{m\,s^{-2}}$ unless indicated otherwise.

Assume that air resistance is negligible and that the ground is level.

1 A projectile attains a certain range when projected at a speed U and angle θ to the horizontal (where its flight starts and ends on the same horizontal level).
At what other angle will the same range be attained, assuming that the speed is unchanged? [3 marks]

2 A particle is projected from the ground at a speed U and angle θ to the horizontal.
Which constant acceleration equations can be used to find

   (i) the maximum height reached by the particle [1 mark]

   (ii) the time taken to reach the maximum height? [1 mark]

3 State two modelling assumptions involving gravity for the motion of a projectile. [2 marks]

4 State a disadvantage of using a particle model for the flight of a golf ball. [2 marks]

5 A particle is projected from ground level and takes T seconds to reach its maximum height, H.
Use two constant acceleration equations to show that it takes a further T seconds to reach the ground again. [4 marks]

6 For the same particle as in question **5**, use a constant acceleration equation to show that its speed on reaching the ground again is the same as its initial speed. [4 marks]

7 A particle is projected from ground level at an angle of 30°, and reaches its maximum height of 10 m after 1 second.
Find its speed of projection. [3 marks]

8 Give two methods for finding the range of a projectile (assuming that the required information is available). [4 marks]

9 A particle is projected from the ground, and the equation of its trajectory is $y = x - 0.1x^2$, where x is measured in metres.
Find

   (i) the angle to the horizontal at which the particle is projected [2 marks]

   (ii) the horizontal distance travelled by the particle before it hits the ground [2 marks]

   (iii) the greatest height reached by the particle [2 marks]

   (iv) the speed of projection of the particle [3 marks]

   (v) the speed of the particle when it hits the ground. [2 marks]

10 A particle is projected downwards at 30° to the horizontal, with a speed of $10\,\mathrm{m\,s^{-1}}$, from a height of 20 m.

   (i) Find the time in flight. [4 marks]

   (ii) Find the speed of the particle when it hits the ground. [7 marks]

11 Assume that $g = 10.0\,\mathrm{m\,s^{-2}}$ for this question.

(i) Show that the Cartesian equation of the trajectory of a projectile, projected with a speed of $u\,\mathrm{m\,s^{-1}}$ at an angle of $\theta°$, is

$$y = x\tan\theta - 5\left(\frac{x}{u\cos\theta}\right)^2 \qquad \text{[4 marks]}$$

A child is attempting to throw a ball over a wall of height 6 m that is 10 m away.

The ball is thrown at a speed of $u\,\mathrm{m\,s^{-1}}$ from a height of 2 m, at an angle θ to the horizontal.

(ii) If the ball just clears the wall, use the Cartesian equation of trajectory to find an expression for u^2 in terms of $\tan\theta$. [3 marks]

(iii) Find u when $\theta = 45°$. [1 mark]

(iv) By differentiating u^2, or otherwise, find the minimum speed at which the ball can be thrown, and the angle at which it must be thrown, in order for the wall to be cleared. [12 marks]

12 A golf ball is struck with a speed of $u\,\mathrm{m\,s^{-1}}$ at an angle of 30° to the ground, and hits the ground again at a distance of R m from the starting position.

It takes T seconds to reach its maximum height. Derive an expression for

(i) T in terms of u and R [3 marks]

(ii) T in terms of u and g [3 marks]

(iii) T in terms of R and g, and hence determine how long the particle is in the air for if it lands 100 m from its starting point. [6 marks]

22 Friction

1 What determines the direction in which a frictional force acts?   [2 marks]

2 If someone is attempting to push an object uphill, but the object does not move, what can be said about the direction of the frictional force? Explain your answer.   [4 marks]

3 What range of values can the coefficient of friction take?   [1 mark]

4 A block rests on a slope which is angled at $\theta°$ to the horizontal. The coefficient of friction between the surface of the slope and the block is μ.

   P_1 is the horizontal force that needs to be applied to the block to stop it from slipping down the slope.

   P_2 is the greatest horizontal force that can be applied without the block slipping up the slope.

   Find expressions for P_1 and P_2.   [15 marks]

5 A sledge of mass 20 kg is being pulled along the ground at constant speed, with a force of PN, by means of a rope that is inclined at 30° to the ground.
   The coefficient of friction between the sledge and the ground is 0.1.
   Other resistances to motion can be ignored.
   The sledge can be treated as a particle.

   (i) Draw a force diagram for the sledge.   [3 marks]

   (ii) Determine P.   [7 marks]

1 Proof

1 **(i)** For example, $a = -2$, $b = 1$

(ii) For example, $a = 1$, $b = -1$

2 **(i)** $A \Leftrightarrow B$, $|x|$ is always positive or zero, so can only equal x when x is positive or zero; thus, the two statements are equivalent.

(ii) $A \Rightarrow B$, John being a pilot implies he has good eyesight, whereas having good eyesight does not imply you are a pilot.

(iii) $A \Rightarrow B$, as the two distinct roots come from a positive discriminant; when the discriminant is zero, there is only one distinct root, so the reverse implication is not true.

3 See worked solution.

4 See worked solution.

5 See worked solution.

6 See worked solution.

7 See worked solution.

8 See worked solution.

9 See worked solution.

10 See worked solution.

2 Trigonometry

1 $14.5°$

2 **(i)** $\dfrac{7}{4}\pi$

(ii) $440°$

3 $16\,\text{cm}$

4 $-\sqrt{3}$

5 $24\,\text{cm}^2$

6 $33.6\,\text{cm}^2$

7 $16.4\,\text{cm}^2$

8 $64.5°, -64.5°, 295.5°$

9 **(i)** See worked solution.

(ii) $-\dfrac{3\pi}{4}$

10 **(i)** $6r - r^2$

(ii) 2

11 **(i)** See worked solution.

(ii) See worked solution.

12 (i) $62.5\,\text{cm}^2$

(ii) $27.3\,\text{cm}^2$

(iii) $17.3\,\text{cm}^2$

13 (i) $\dfrac{\sqrt{3}}{2}r^2$

(ii) $\dfrac{3\sqrt{3}}{2} < \pi < 2\sqrt{3}$

14 (i) $40.2\,\text{cm}$

(ii) $73.9\,\text{cm}^2$

3 Sequences and series

1 $-3, -10, -17, -24, -31$

2 $u_n = -3.5 + 1.5n$

3 $1, 0.5, 0.25, 0.125, 0.0625$

4 $u_n = \dfrac{1}{4} \times 2^n$

5 See worked solution.

6 $|r| < 1$

7 See worked solution.

8 Eight terms are needed.

9 (i) See worked solution.

(ii) $22.5\dot{1}$

10 (i) See worked solution.

(ii) $n + 1$ pieces

11 (i) $a = \dfrac{-(n-1)d}{2}$

(ii) (a) Not possible, as n cannot be negative (it is the number of terms).

(b) Possible as $n - 1$ will end up positive.

(c) Possible as $n - 1$ will be even and positive.

(d) Only possible if d is divisible by two, since $n - 1$ will be odd.

12 (i) See worked solution.

(ii) $\dfrac{8}{27}$

13 4

14 $r = 3 \pm 2\sqrt{2}$

4 Functions

1 A is many-to-one and is also a function.

B is many-to-many and is **not** a function.

C is one-to-many and is **not** a function.

2 $f(x) \geqslant -5$

3

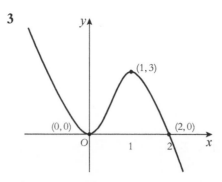

Passes through (0,0), (1,3) and (2,0).

4 $9x^2 - 42x + 50$

5 $x = \pm\sqrt{6}$

6 $|x - 3.5| < 1.5, a = 3.5$ and $b = 1.5$

7

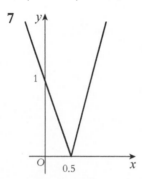

8 **(i)** $f(x) \geqslant 3$

   (ii) $g(x) \geqslant 5$

   (iii) $0 < x < 0.625$

9 **(i)** **(a)** $x^2(2x + 3)$

      (b) $-2x^3 - 3x^2$

   (ii) **(a)** Function is many-to-one (graph goes up and down), so no inverse.

      (b) $x > 1.5$, for example.

10 (i) **(a)** Points (2,3), (4,1), (7,–2) are on $y = f(x - 2)$.

     (b) Points (0,9), (2,3), (5,–6) are on $y = 3f(x)$.

     (c) Points (0,2), (2,0), (5,–3) are on $y = f(x) - 1$.

     (d) Points (0,3), (1,1), (2.5,–2) are on $y = f(2x)$.

  (ii) **(a)** $f^{-1}(x)$ does not exist (unless the domain of $f(x)$ is restricted).

     (b) If $f^{-1}(x)$ exists then points (3,0), (1,2), (–2,5) are on $y = f^{-1}(x)$.

11 (i) **(a)** 67

     (b) $fgh(x) > 22$

  (ii) **(a)** $fg(x) = 15x^2 + 7$, $gf(x) = 75x^2 + 60x + 13$

     (b) $x = \dfrac{-10 \pm \sqrt{60}}{20} = -0.113, -0.887$

12 (i) Curve crosses y-axis at $(0, \frac{1}{k})$, and x-axis at $(a, 0)$.

(ii) $f^{-1}(x) = \dfrac{a(kx - 1)}{x - 1}$

(iii) $x = 0$ or $\dfrac{5}{3}$

13 (i)

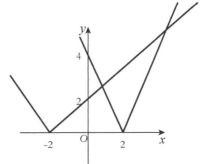

(ii) $\dfrac{2}{3} < x < 6$

5 Differentiation

1 $\dfrac{d^2 y}{dx^2} = 0$

2 It is a maximum.

3 $\dfrac{dy}{dx} = 0$ and $\dfrac{d^2 y}{dx^2} = 0$

4 $-2 \leqslant x \leqslant 2$

5 $\dfrac{dy}{dx} = \dfrac{3 - \dfrac{1}{x^2}}{2\sqrt{3x + \dfrac{1}{x}}}$

6 $\dfrac{dy}{dx} = \dfrac{3x - 10}{2\sqrt{x - 5}}$

7 $\dfrac{dy}{dx} = \dfrac{-x^2 - 2}{\left(x^2 + 3x - 2\right)^2}$

8 $(-1, -12)$, $(2, -45)$

9 (i) $x < 0$

(ii) $(0, 10)$

10 (i) $0.000\,955\,\mathrm{cm\,s^{-1}}$

(ii) The shape of the cone may need to change, with radius to height ratio increasing.

11 $\dfrac{3}{7}$

12 $\dfrac{3}{2\pi}\,\mathrm{m\,min^{-1}}$

13 $\left(-\dfrac{1}{2}, -1\right)$

14 (i) Radius decreases at $\dfrac{1}{200\pi}$ cm s^{-1}

   (ii) Surface area decreases at $\dfrac{2}{5}$ cm^2 s^{-1}

15 See worked solution.

16 (i) See worked solution.

   (ii) $(1,1)$ and $(-3,1)$

   (iii) $PQ = 4$

6 Trigonometric functions

1 0.4 rad

2 112.5°

3 Five solutions.

4 2

5 $\pm\dfrac{\pi}{3}$

6 $\sqrt{15}$

7 $-\dfrac{\pi}{2}, \dfrac{\pi}{2}, \dfrac{\pi}{6}, \dfrac{5\pi}{6}$

8 0 or $\dfrac{\pi}{3}$

9 $-\dfrac{4\pi}{9}, \dfrac{2\pi}{9}, \dfrac{8\pi}{9}$

10 (i)

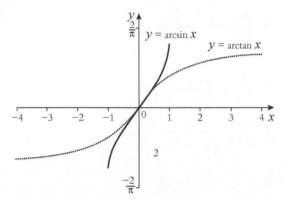

   (ii) (a) $\arcsin(1) = \dfrac{\pi}{2}, \arctan(1) = \dfrac{\pi}{4}$

   (b) See worked solution.

11 $\dfrac{1 + \sqrt{5}}{2}$

12 (i) 144 m (3 s.f.)

   (ii) (a) 94.1 m

   (b) 121 m (3 s.f.)

13 (i) See worked solution.

   (ii) $\theta = 0, 2\pi$

14 (i) See worked solution.

  (ii) $-\dfrac{7\pi}{16}, -\dfrac{3\pi}{16}, \dfrac{\pi}{16}, \dfrac{5\pi}{16}$

  (iii) Three solutions.

7 Further algebra

1 $|x| < \dfrac{1}{5}$

2 See worked solution.

3 $x^2 + 4x + 5 + \dfrac{5}{x-1}$

4 $1 - 2x + 3x^2 - 4x^3$

5 $\dfrac{x(x+1)}{2x^3 - 1}$

6 $\dfrac{19\,799}{14\,000}$

7 $1 + x - 3x^2 + 5x^3$

  Valid for $|x| < 1$

8 $\dfrac{0.51}{1-3x} + \dfrac{0.17}{3+x} - \dfrac{0.1}{(3+x)^2}$

9 (i) $1 + \dfrac{2}{3}x - \dfrac{4}{9}x^2$

  (ii) $|x| < \dfrac{1}{2}$

10 $2x - 1 + \dfrac{4}{2x-1} - \dfrac{1}{x+1}$

11 $\dfrac{1}{2} + \dfrac{1}{48}x + \dfrac{1}{576}x^2$ for $|x| < 8$

12 (i) $\dfrac{7x+5}{(x-1)(x+1)}$

  (ii) $-5 - 7x - 5x^2 - 7x^3$

  (iii) 0.01%

13 (i) $6 + 5x - 7x^2$

  (ii) $|x| < 1$

  (iii) $\dfrac{\sqrt{7}}{7}$ or 0.378 (3 s.f.)

14 (i) $\dfrac{4x+5}{(1+5x)(1-x)} = \dfrac{7}{2(1+5x)} + \dfrac{3}{2(1-x)}$

  (ii) $5 - 16x + 89x^2 - 436x^3$

  (iii) $|x| < \dfrac{1}{5}$

8 Trigonometric identities

1 See worked solution.

2 See worked solution.

3 (α, R)

4 Translation of $y = \cos x$ by $\begin{pmatrix} \alpha \\ 0 \end{pmatrix}$.

Stretch of $y = \cos(x - \alpha)$ by a factor R parallel to the y-axis.

5 $\theta = 0, \dfrac{\pi}{2}, \pi$

6 $\dfrac{\sqrt{2}}{2}$

7 See worked solution.

8 (i) $\sqrt{2}\cos(x + 45°)$

  (ii) $-\sqrt{2}$

  (iii) $40.9°$ or $229.1°$

9 (i) $\sqrt{13}\sin(x - 0.983)$

  (ii) $\dfrac{1}{k + \sqrt{13}}$

  (iii) $2 - \sqrt{13}$

10 $-a$

11 $\tan 15° = 2 - \sqrt{3}$

12 $0, \ \pi, \ 2\pi, \ \dfrac{7\pi}{6}, \ \dfrac{11\pi}{6}$

13 (i) $2\cos(x - 30°)$

  (ii) $x = 75°, \ 345°$

14 $\theta = -\dfrac{\pi}{2}, \ -\dfrac{\pi}{6}, \dfrac{\pi}{2}, \ \dfrac{5\pi}{6}$

9 Further differentiation

1 See worked solution.

2 $k = 3$

3 2

4 See worked solution.

5 $\dfrac{dy}{dx} = \dfrac{ax}{by}$

6 $\dfrac{1}{e}$

7 $\dfrac{3}{1 + 9x^2}$

8 $\dfrac{1}{\sqrt{1 - x^2}}$

9 (i) See worked solution.

(ii) $x = \dfrac{\pi}{3}, \dfrac{5\pi}{3}$

10 $-\dfrac{1}{e}$

11 (i) $y = -\dfrac{1}{2}x + 2$, $y = \dfrac{1}{2}x - 2$

(ii) Intersection point $(4,0)$.

12 (i) $y = \dfrac{b}{a}x$

(ii) See worked solution.

13 $24\ln 4 + 9$

14 See worked solution.

10 Integration

1 $\dfrac{5}{3}\left(e^{3} - 1\right)$

2 $\dfrac{1}{3}\sin 3x - \dfrac{2}{3}\cos 3x + c$

3 $\dfrac{1}{2}\sqrt{3}$

4 $t^{2} - \dfrac{9}{5}t^{\frac{5}{3}} + c$

5 $-\dfrac{1}{2}\ln\left(5 - 2t\right) + c$

6 See worked solution.

7 $f\left(x\right) = \dfrac{1}{10}\left(3 + x^{2}\right)^{5} - \dfrac{3}{10}$

8 $\ln\left(at^{2}\left(t + 3\right)\right)$

9 See worked solution.

10 4

11 $\dfrac{1}{2}\left(\ln x\right)^{2} + c$

12 See worked solution.

13 $\dfrac{320}{21} + \dfrac{128}{21}\sqrt{2}$

14 $\dfrac{1}{2}e^{2x} + 4e^{x} + 4x + c$

15 $3x - \dfrac{3}{4}\sin 4x + c$

16 (i) $\dfrac{5}{2}\left(e^{4} - 1\right)$

(ii) $\dfrac{5}{4}\left(3e^{4} + 1\right)$

17 6

18 $\dfrac{1}{2}\ln 2$

19 (i) See worked solution.

   (ii) $\dfrac{2}{5}$

20 $\ln\left(\dfrac{(x+1)}{(3x+1)^{\frac{2}{3}}}\right) - \dfrac{1}{3(3x+1)} + c$

21 (i) $\ln(\ln x) + c$

   (ii) $\ln 2$

22 $\pi^2 - 4$

11 Parametric equations

1 $y^2 = 12x$

2 $\dfrac{1}{t}$

3 $y = \dfrac{4}{x}$

4 $x = 2\cos\theta,\ y = 2\sin\theta$

5 $x = 1 + \cos\theta,\ y = 2 + \sin\theta$

6 $\dfrac{\pi}{2}$ and $\dfrac{3\pi}{2}$

7 $-\dfrac{\pi}{4}$ and $\dfrac{3\pi}{4}$

8 The domain of x is limited to $x \geqslant 0$.

9 (i) $y^3 + y^2 = 8x$

   (ii) $\left(\dfrac{14}{27}, \dfrac{4}{3}\right)$ and $\left(-\dfrac{1}{2}, -2\right)$

   (iii) $\dfrac{\mathrm{d}y}{\mathrm{d}x} = \dfrac{2}{3t^2 + t}$

   (iv) $t = \dfrac{2}{3}$ or -1

10 (i) $9\sin^2 3t + 4\cos^2 2t$

   (ii) See worked solution.

11 (i) $x^2 + 1 = \dfrac{1}{y^2}$

   (ii) $(0,1)$ and $(0,-1)$

12 $\dfrac{1}{e^{\pi}}$

13 (i) $\dfrac{x^2}{9} + \dfrac{y^2}{4} = 1$

   (ii) $(3, 0), (-3, 0), \left(\dfrac{9}{5}, \dfrac{8}{5}\right)$ and $\left(\dfrac{9}{5}, -\dfrac{8}{5}\right)$

14 $\dfrac{x^2}{4} - \dfrac{y^2}{4} = 1$

15 $y = \dfrac{3x^2}{1 + x}$

12 Vectors

1 26, 67.4° (3 s.f.) to the positive x-axis.

2 $-5\sqrt{3}\mathbf{i} + 5\mathbf{j}$ N

3 3

4 $9\sqrt{2}$

5 $-\dfrac{4}{5\sqrt{2}}\mathbf{i} + \dfrac{3}{5\sqrt{2}}\mathbf{j} - \dfrac{1}{\sqrt{2}}\mathbf{k}$

6 See worked solution.

7 28.0 N (3 s.f.), 30.4° (3 s.f.) to the x-axis.

8 $a = 2$ and $b = -2$

9 $\mathbf{d} = 2\mathbf{b}, \mathbf{c} = \mathbf{b} - \mathbf{a}$

10 $(-1.4, 3.6)$

13 Differential equations

1 (i) One where there are no derivatives higher than the first.

   (ii) The constant of integration is left in the solution, so it refers to a family of curves.

   (iii) Additional information is used to find the constant of integration, so it refers to one curve.

2 See worked solution.

3 $y = x^2 - x + c$

4 $y = Ae^{\frac{1}{2}x^2}$

5 $y = x^2 - x + 1$

6 $y = e^{\frac{1}{2}x^2 - 2}$

7 $y = \dfrac{1}{2}\left(5e^{2x} + 1\right)$

8 (i) $\dfrac{dv}{dt} = \dfrac{k}{v}$

   (ii) $\dfrac{dv}{dt} = \dfrac{20}{v}$

   (iii) $v = \sqrt{40t + c}$

   (iv) $v = \sqrt{40t}$

   (v) $t = 10$ s

9 $\dfrac{dT}{dt} = \dfrac{1}{320}(175 - T)$

10 21 billion

11 $y = \dfrac{2x^2}{1 + x^2}$

14 Numerical methods

1 See worked solution.

2 1 ± 0.1

3 Discontinuity at $\sqrt{2}$, masks root.

4 Overestimate, concave upwards.

5 Lower bound 12.52; upper bound 17.60

6 Negative root cannot be found from an integer starting value.

7 **(i)** See worked solution.

   (ii) (a) $x = 1$

       (b) =B2

       (c) 1.2056

8 **(i)** 0.5

   (ii) =0.5*A$2*(C4+C5)

   (iii) 7.4375

9 Correct, but not justified as no check for change in sign.

10 **(i)** See worked solution.

   (ii) 0.106

   (iii) $[-0.1, 0]$ and $[0.7, 0.8]$

   (iv) There are two roots in $[0,1]$, so decimal search fails to spot them.

11 **(i)** See worked solution.

   (ii) See worked solution.

   (iii) 0.57

   (iv)

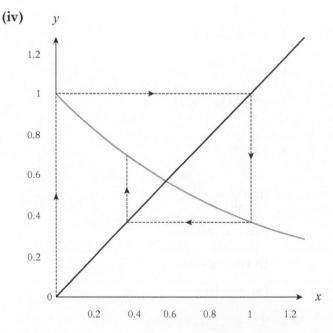

   (v) The iterations form a cobweb and the gradient of g(x) is small and so the convergence is slow.

   (vi) Only by one step, as 1 is the second value.

12 (i) 1.458

  (ii)

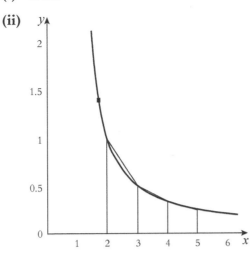

Overestimate

13 (i) See worked solution.

  (ii) See worked solution.

  (iii) $-1.39, 0.33, 1.21$

14 (i) $109.6\,\text{m}^2$

  (ii) $138.6\,\text{m}^2$

  (iii) Second, as more strips.

15 (i) See worked solution; [4,5].

  (ii) See worked solution.

  (iii) 0.7 and 4.3

  (iv)

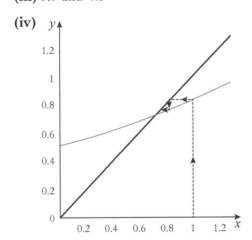

  (v) Starting with 4 for the rearrangement in ① results in the root in [0,1]. Starting with 5 for the rearrangement in ① results in divergent iterations.

Starting with 1 for the rearrangement in ② results in the root in [4,5]. Starting with 1 for the rearrangement in ② results in an error as ln is undefined at 0.

16 $-0.54, 1.49, 2.62$

17 (i) LB $= 2.3926$, UB $= 6.0937$

  (ii) $4.2431\,\text{m}$

  (iii) Use more strips.

18 (i) See worked solution.

(ii) 0.72

(iii)

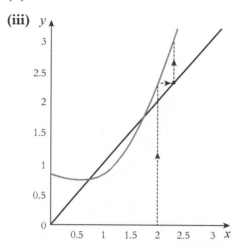

Iterations diverge from the root.

(iv) 1.70

15 Probability

1 See worked solution.

2 0.88

3 0.65

4 (i) $\dfrac{2p^2}{3}$

(ii) $\dfrac{p}{3}(9 - 2p)$

5 Independent but not mutually exclusive.

6 0.58

7 French student being male is more likely $(0.5939 > 0.5765)$.

8 (i)

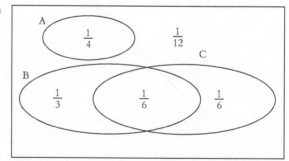

(ii) $\dfrac{1}{4}$

9 (i)

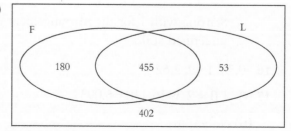

(ii) (a) $\dfrac{455}{1090}$

(b) $\dfrac{455}{1090}$

(c) $\dfrac{455}{508}$

(d) $\dfrac{455}{635}$

(iii) $P(F \cap L) = 0.4174$

$P(F) \times P(L) = 0.2715$

So, the events are not independent.

10 (i) 0.375

(ii) 45.75

(iii) $\dfrac{36}{61}$

11 (i)

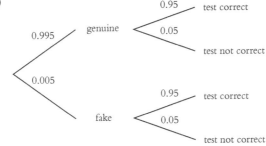

(ii) 0.0545

(iii) $\dfrac{19}{218}$

16 Statistical distributions

1 (i) $\sqrt{10}$

(ii) (a) 0.0007827

(b) 0.9431

(c) 0.7717

2 (i) 0.006210

(ii) 0.04779

3 39.8 (3 s.f.)

4 75.03 mph

5 111.5

6 3.947

7

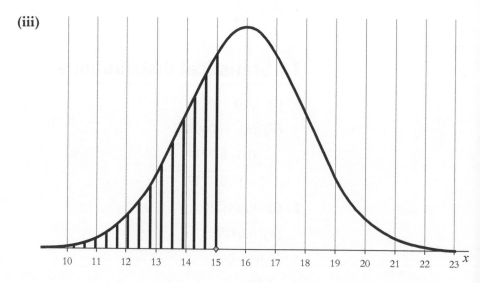

8 (i) $Y \sim N(14, 3.6^2)$

   (ii) 0.8667

9 3 gold; 16 silver; 1 'I tried'

10 (i) 0.04006

   (ii) 6.428×10^{-5}

   (iii) 1.6

11 (i) 2.75 kg

   (ii) 22.3 kg

12 (i) Most of the population lies within three standard deviations from the mean, so the range is approximately six standard deviations.

   (ii) 16 cm

   (iii)

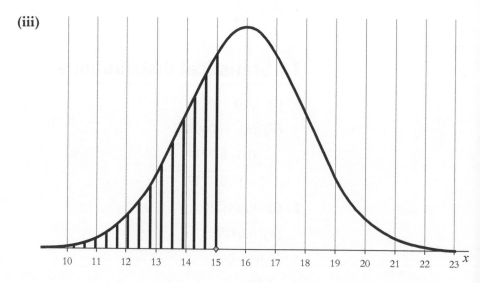

   (iv) 0.31 (0.3085 to 4 s.f.)

13 (i) 200.822 cm

   (ii) 0.92 (2 d.p.)

14 (i) $\sigma = 4292$ km

   (ii) 0.0649

15 (i) $\mu = 63, \sigma = 14.83$

   (ii) $X > 92.66$ or $X < 33.34$

16 (i) See worked solution.

(ii) $159.5 = \mu + 0.8416\sigma$

(iii) $\mu = 87.44, \sigma = 85.63$

(iv) See worked solution.

(v) Not normal distribution.

17 Hypothesis testing

1 Non-linear graph.

2 Association.

3 (i) Points close to a curve.

(ii) H_0 : There is no association between BSA and weight.

H_1 : There is some association between BSA and weight.
Reject null hypothesis.

4 (i) H_0 : There is no association between age and time spent watching television.

H_1 : There is some association between age and time spent watching television.

(ii) Accept null hypothesis.

5 (i) Ages all integers.

(ii) Non-linear.

(iii) Reject null hypothesis.

6 (i) Weekdays and weekends.

(ii) Reject null hypothesis.

(iii) Maria uses association and not correlation which is not correct when correlation coefficient is used. Maria uses the word proof; however, she has made an *inference* not a proof.

7 $H_1 : \mu \neq 14$ 2-tail, $H_1 : \mu < 14$ 1-tail, $H_1 : \mu > 14$ 1-tail

8 $\bar{X} \sim N\left(\mu, \dfrac{\sigma^2}{n}\right)$

9 (i) $\bar{X} \sim N(50, 0.4)$

(ii) (a) $\bar{X} > 50.8105244$

(b) $\bar{X} < 49.1894756$

(c) Either $\bar{X} < 48.9597032$ or $\bar{X} > 51.0402968$

10 (i) 0.1056

(ii) 0.0062

(iii) 0.0304

11 (i) $\bar{X} \sim N(13.5, 0.45)$

(ii) $H_0 : \mu = 13.5, H_1 : \mu < 13.5$

(iii) 0.0680

(iv) Not enough evidence to reject the null hypothesis.

12 (i) $H_0 : \mu = 130, H_1 : \mu < 130$

(ii) $\bar{X} < 124.798$

(iii) Not enough evidence to reject the null hypothesis.

13 (i) $H_1 : \mu \neq 17$

(ii) $\bar{X} < 16.30704808$ or $\bar{X} > 17.69295192$

(iii) Not enough evidence to reject the null hypothesis

14 (i) $H_0 : \mu = 21, H_1 : \mu \neq 21$

(ii) $\bar{X} \sim N(21, 0.9)$

(iii) Enough evidence to reject the null hypothesis.

15 Not enough evidence to reject the null hypothesis.

18 Kinematics

1 $7\,\text{m s}^{-1}$

2 $t = 15\,\text{seconds}$

3 $3.46\,\text{seconds}$

4 (i) $X = 3.27, Y = 12.61$

(ii) $y = 12.61 + 0.15x - 1.225x^2$

(iii) $12.4\,\text{m}$ above the sea

5 (i) $11\mathbf{i} + 5\mathbf{j}$

(ii) $T = 50\,\text{minutes and}\,24\,\text{seconds after midday.}$

6 $x = 15.5$ and $y = 3.972$

19 Forces and motion

1 $3.47\,\text{N}$

2

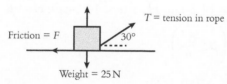

3 $-8\mathbf{i} + 4\mathbf{j}$

4 $\sqrt{65} = 8.06\,\text{N}$, at $119.7°$

5 $\begin{pmatrix} 26 \\ 8 \end{pmatrix}$

6 $6.61\,\text{N}$

7 A: $T = 2a$; B: $g - T = a$

8 (i)

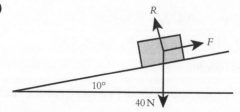

   (ii) 6.95 N

   (iii) 20.0 N

9 (i) 0.8

   (ii) 200 N

10 (i) See worked solution.

   (ii) (a) cos BAC = 0.6

   (b) cos BCA = 0.8

   (iii) (a) 192 N

   (b) 144 N

11 (i) $T = 2000\,\text{N}$

   (ii) $F = 2800\,\text{N}$

   (iii) 3730 N

12 (i) $x = -6$

   (ii) $m = 3$

13 (i) $a = 0.00333\,\text{m s}^{-2}$

   (ii) 2880 m

14 (i) 139.5°

   (ii) 2.84P

   (iii) 138.7°

20 Moments

1 If we take moments about B, we obtain $dX + 2dZ$, which cannot equal zero. Thus, the system cannot be in equilibrium.

2 The total forces on the object in two perpendicular directions will be zero. This gives two equations. The total moment of the forces (about any point) will be zero, giving a third equation.

3 The net moment will be the same for all points. Choose the most convenient point: one at which an unknown force is applied, or such that the least number of forces appears in the expression for the net moment.

4 $X = 20$; $d = 0.25$

5 See worked solution.

6 $\frac{8}{3}$ m from A

21 Projectiles

1 $90 - \theta$

2 (i) $v^2 = u^2 + 2as$

   (ii) $v = u + at$

3 The acceleration due to gravity is constant (the variation in vertical distance travelled is relatively small). The acceleration due to gravity has a constant direction (the horizontal distance travelled is relatively small).

4 A particle does not rotate, but a golf ball usually does.

5 See worked solution.

6 See worked solution.

7 $40\,\mathrm{m\,s^{-1}}$

8 Find the time in flight and multiply by the horizontal component of the initial velocity. Set the vertical displacement equal to zero in the Cartesian equation of the trajectory.

9 (i) $\theta = 45°$

  (ii) $10\,\mathrm{m}$

  (iii) $2.5\,\mathrm{m}$

  (iv) $9.90\,\mathrm{m\,s^{-1}}$

  (v) $9.90\,\mathrm{m\,s^{-1}}$

10 (i) $1.57\,\mathrm{s}$ (3 s.f.)

  (ii) $22.2\,\mathrm{m\,s^{-1}}$ (3 s.f.)

11 (i) See worked solution.

  (ii) $u^2 = \dfrac{250\left(\tan^2\theta + 1\right)}{5\tan\theta - 2}$

  (iii) $12.9\,\mathrm{m\,s^{-1}}$

  (iv) $12.2\,\mathrm{m\,s^{-1}}$ (3 s.f.), $55.9°$

12 (i) $T = \dfrac{R}{u\sqrt{3}}$

  (ii) $T = \dfrac{u}{2g}$

  (iii) $3.43\,\mathrm{s}$

22 Friction

1 Friction opposes motion or attempted motion.

2 It could be in either direction, or the frictional force could be zero.
There are potentially three forces at work along the slope: the pushing force, the downhill component of the weight of the object, and the frictional force.

At one extreme, the object is on the point of moving uphill, and the frictional force will act downhill. At the other extreme, the object is on the point of moving downhill, and the frictional force will act uphill.

If the pushing force balances the downhill component of the weight of the object, then there will be no frictional force.

3 $\mu > 0$

4 $P_1 = \dfrac{mg\left(\sin\theta - \mu\cos\theta\right)}{\cos\theta + \mu\sin\theta}$, $P_2 = \dfrac{mg\left(\sin\theta + \mu\cos\theta\right)}{\cos\theta - \mu\sin\theta}$

5 (i)

  (ii) $21.4\,\mathrm{N}$